Nuggets of Number Theory

A Visual Approach

AMS / MAA | CLASSROOM RESOURCE MATERIALS

VOL 55

Nuggets of Number Theory

A Visual Approach

Roger B. Nelsen

Providence, Rhode Island

2010 *Mathematics Subject Classification.* Primary 11-01;
Secondary 11A07, 11B39, 11D04.

For additional information and updates on this book, visit
www.ams.org/bookpages/clrm-55

Library of Congress Cataloging-in-Publication Data

Names: Nelsen, Roger B., author.
Title: Nuggets of number theory : a visual approach / Roger B. Nelsen.
Description: Providence, Rhode Island : MAA Press, an imprint of the American Mathematical Society, [2018] | Series: Classroom resource materials ; volume 55 | Includes bibliographical references and index.
Identifiers: LCCN 2018000043 | ISBN 9781470443986 (alk. paper)
Subjects: LCSH: Number theory–Study and teaching. | Mathematics–Study and teaching. | AMS: Number theory – Instructional exposition (textbooks, tutorial papers, etc.). msc | Number theory – Elementary number theory – Congruences; primitive roots; residue systems. msc | Number theory – Sequences and sets – Fibonacci and Lucas numbers and polynomials and generalizations. msc | Number theory – Diophantine equations – Linear equations. msc
Classification: LCC QA241 .N435 2018 | DDC 512.7–dc23
LC record available at https://lccn.loc.gov/2018000043

∞ The paper used in this book is acid-free and falls within the guidelines
established to ensure permanence and durability.
Visit the AMS home page at `http://www.ams.org/`

10 9 8 7 6 5 4 3 2 1 23 22 21 20 19 18

Contents

Preface

The theory of numbers is the last great uncivilized continent of mathematics. It is split up into innumerable countries, fertile enough in themselves, but all the more or less indifferent to one another's welfare and without a vestige of a central, intelligent government. If any young Alexander is weeping for a new world to conquer, it lies before him.

Eric Temple Bell

The elementary theory of numbers should be one of the very best subjects for early mathematical instruction. It demands very little previous knowledge; its subject matter is tangible and familiar; the processes of reasoning which it employs are simple, general, and few; and it is unique among the mathematical sciences in its appeal to natural human curiosity. A month's intelligent instruction in the theory of numbers ought to be twice as instructive, twice as useful, and at least ten times as entertaining as the same amount of "calculus for engineers."

Godfrey Harold Hardy

Number theorists are like lotus-eaters—having once tasted of this food they can never give it up.

Leopold Kronecker

Some time ago I was looking at several textbooks for the undergraduate number theory course. I was struck by how few illustrations were included in many of those textbooks. A number—specifically a positive integer—can represent many things: the cardinality of a set; the length of a line segment; or the area of a plane region. Such representations naturally lead to a variety of visual arguments for topics in elementary number theory. Since the number theory course usually begins with properties of the positive integers, the texts should have more pictures. That observation became the motivation for this book.

Work on this book began when I was invited to give a talk at the MAA's MathFest in Albuquerque in August 2005, in a session entitled "Gems of Number Theory" organized by Arthur Benjamin and Ezra

Brown. The title of that talk was "Some visual gems from elementary number theory." Later a version of the talk appeared as an article in the February 2008 issue of *Math Horizons*. This article was subsequently included in the 2009 MAA book *Biscuits of Number Theory*, edited by Benjamin and Brown.

Nuggets of Number Theory: A Visual Approach is not a textbook. Although it is designed to be used by the instructor of an undergraduate number theory course as a supplement to a standard textbook, it will be of interest to anyone who loves number theory. Certain chapters in the book may be of interest to those who teach discrete mathematics, abstract algebra, and teacher preparation courses. The book can also be used as a resource for group projects or extra-credit assignments.

A *nugget* is a lump of precious metal or, more generally, something of significance or of great value. The nuggets in this book are topics in number theory for which I believe a visual approach is appropriate and beneficial, with a chapter devoted to each one. Chapter 1 is devoted to figurate numbers—numbers that can be represented by objects such as pebbles arranged in geometrical patterns—that were studied by the early Greeks. Chapter 2 deals with the important concept of congruence, and includes visual demonstrations of Fermat's little theorem and Wilson's theorem. Visual approaches to the solutions to linear Diophantine equations and Pell equations are the subject of Chapter 3. Pythagorean triples—solutions to the Diophantine equation $a^2 + b^2 = c^2$—are represented naturally by integer-sided right triangles and explored in Chapter 4. Certain irrational numbers and their rational approximations, including ones based on continued fractions, appear in Chapter 5. Many identities for the Fibonacci and Lucas numbers are presented visually in Chapter 6, and similarly for properties of perfect numbers in Chapter 7. Each chapter includes a set of exercises, with solutions to all the exercises following the final chapter. The book concludes with references and a complete index.

Acknowledgments. I would like to express my appreciation and gratitude to Susan Staples and the members of the editorial board of Classroom Resource Materials for their careful reading of an earlier draft of this book, and for their many helpful suggestions. I would also like to thank Stephen Kennedy, Carol Baxter, Beverly Ruedi, and Bonnie Ponce of the MAA publication staff and Christine Thivierge, Sergei Gelfand, Courtney Rose, and Jennifer Wright Sharp of the AMS production staff for their encouragement, expertise, and hard work in preparing this book for publication.

Roger B. Nelsen
Lewis & Clark College
Portland, Oregon

CHAPTER 1

Figurate Numbers

Why are numbers beautiful? It's like asking why is Ludwig van Beethoven's Ninth Symphony beautiful. If you don't see why, someone can't tell you. I know numbers are beautiful. If they aren't beautiful, nothing is.

Paul Erdős

Mighty are numbers, joined with art resistless.

Euripides

The *figurate numbers* are positive integers that can be represented geometrically by arrangements of points, or physically by arrangements of objects like pebbles. As such they form a connection between geometry and number theory—indeed, number words such as *square* and *cube* reflect this relationship. Figurate numbers appear in some of the works by early Greek geometers, and today they regularly appear in undergraduate number theory texts. In fact, figurate numbers appear in every chapter in this book. Our approach to the figurate numbers reflects their origin in geometry.

We begin with plane figurate numbers known as polygonal numbers, represented by arrangements of objects in the plane in polygonal patterns. We conclude this chapter with several three-dimensional figurate numbers.

1.1. Polygonal numbers

The *triangular numbers* are sums of consecutive positive integers, e.g., 1, 1+2, 1+2+3, etc., and can be represented by arrangements of objects in a triangle, as shown in Figure 1.1(a). We use the notation $T_k = 1 + 2 + \cdots + k$ for the kth triangular number for $k \geq 1$ (and define $T_0 = 0$). The shape of the triangle is arbitrary, usually it is drawn as an equilateral, isosceles, or right triangle. Similarly, the squares are, of course, represented by objects in a square pattern, as shown in Figure 1.1(b).

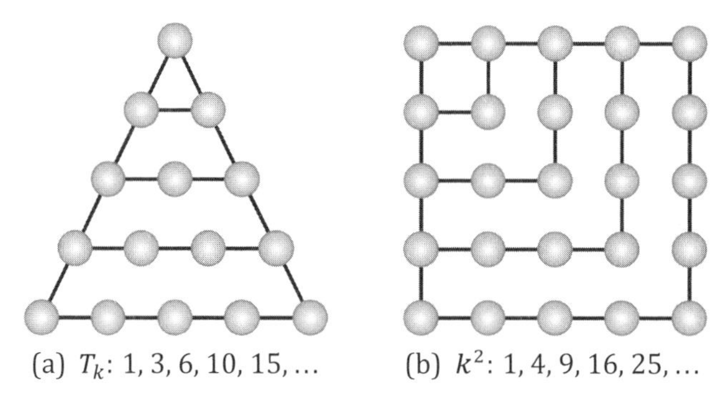

(a) T_k: 1, 3, 6, 10, 15, ... (b) k^2: 1, 4, 9, 16, 25, ...

FIGURE 1.1

Observe in Figure1.1(b) that $k^2 = 1 + 3 + 5 + \cdots + (2k - 1)$, that is, k^2 is the sum of the first k odd positive integers, analogous to $T_k = 1 + 2 + \cdots + k$ for the triangular numbers.

Raphael and the fourth triangular number

In his classic painting *The School of Athens*, the Italian renaissance painter Raphael (1483–1520) included Pythagoras in the lower left corner, writing in a book, with a slate at his feet. At the bottom of the slate is the *tetractys*, the representation of T_4 as ten objects (Roman numerals) arranged in a triangular pattern, with the sum X.

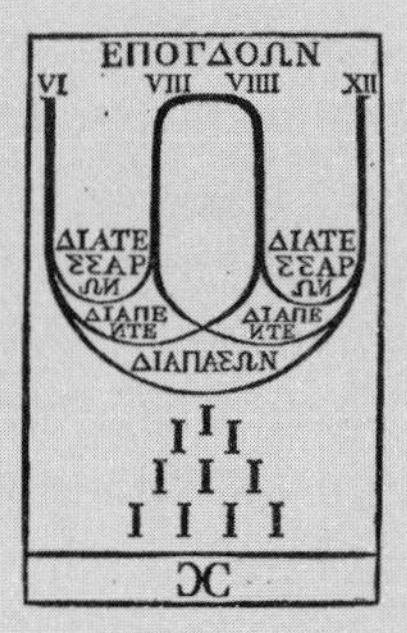

The School of Athens, Pythagoras, and the *tetractys*

The drawing of the slate is from the book *Descrizione delle immagini dipinte da Raffaelle d'Urbino* by Giovanni Bellori, published in 1751.

Carl Friedrich Gauss and the 100th triangular number

Nearly every biography of the great mathematician Carl Friedrich Gauss (1777–1855) relates the following story. When Gauss was about ten years old, his arithmetic teacher asked the students in class to compute the sum

$1 + 2 + 3 + \cdots + 100$, anticipating this would keep them busy for some time. He barely finished stating the problem when young Carl came forward and placed his slate on the teacher's desk, void of calculation, with the correct answer: 5050. When asked to explain, Gauss admitted he recognized the pattern $1+100 = 101, 2+99 = 101, 3+98 = 101$, and so on to $50+51 = 101$. Since there are fifty such pairs, the sum must be $50\cdot 101 = 5050$. The pattern for the sum (adding the largest number to the smallest, the second largest to the second smallest, and so on) is illustrated below, along with a portrait of Gauss on a pre-euro 10 Deutsche Mark note.

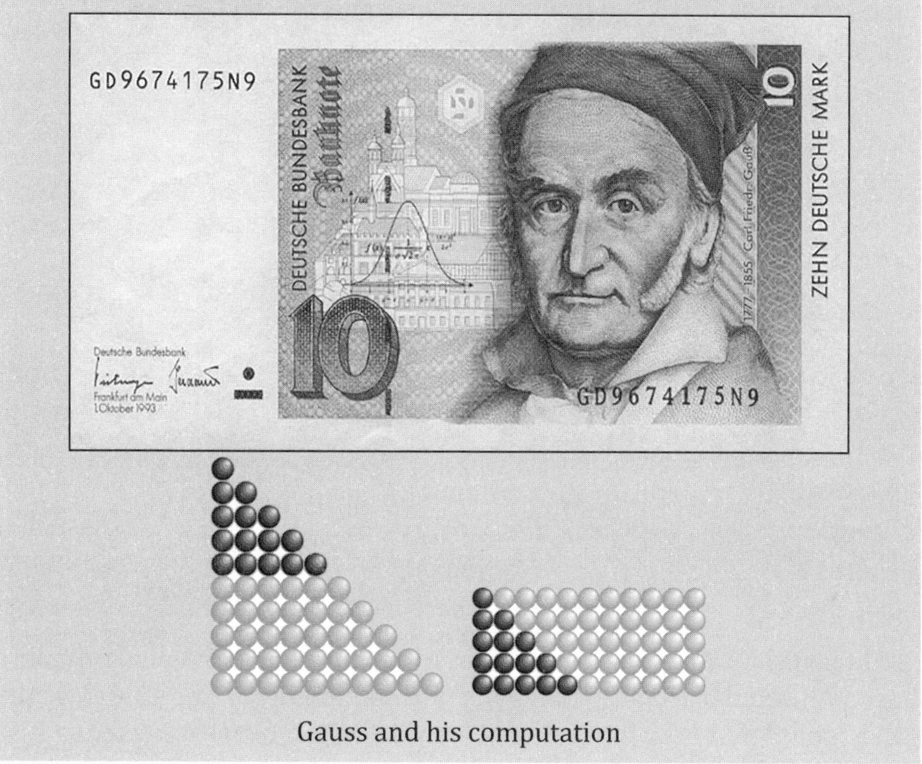

Gauss and his computation

Two more types of polygonal numbers are the *oblong numbers* O_k (objects in a rectangular array with k rows and $k + 1$ columns) and the *pentagonal numbers* P_k (objects in a pentagonal pattern), as shown in Figure 1.2.

Clearly $O_k = k(k + 1)$, and $O_k = 2 + 4 + 6 + \cdots + 2k$, that is, O_k is the sum of the first k even positive integers.

In Figure 1.3 we illustrate two simple but important relationships involving the triangular, square, and oblong numbers.

Figure 1.3(a) yields a formula for the triangular numbers $T_k = \frac{k(k+1)}{2}$, while Figure 1.3(b) foreshadows a method we shall use to find formulas for other polygonal numbers.

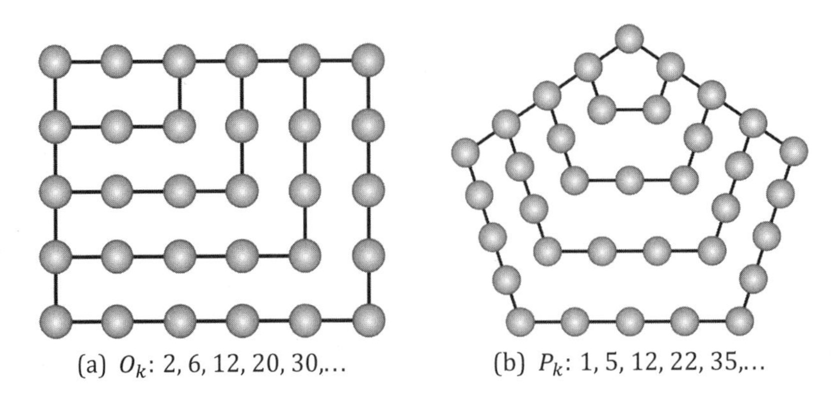

(a) O_k: 2, 6, 12, 20, 30,... (b) P_k: 1, 5, 12, 22, 35,...

FIGURE 1.2

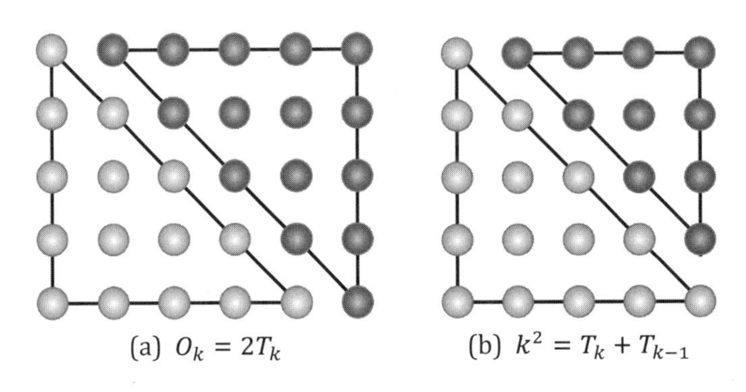

(a) $O_k = 2T_k$ (b) $k^2 = T_k + T_{k-1}$

FIGURE 1.3

In Table 1.1 we exhibit the values of the first few polygonal numbers based on regular *n-gons* (polygons with n sides), i.e., the values of the kth *n-gonal numbers* for $3 \le n \le 8$ and $1 \le k \le 10$. We refer to k as the *rank* of the n-gonal number.

Table 1.1 yields some important information about a general formula for the kth n-gonal number. The formula should be linear in n,

TABLE 1.1

$k =$	1	2	3	4	5	6	7	8	9	10
$n = 3$ (Triangular)	1	3	6	10	15	21	28	36	45	55
$n = 4$ (Square)	1	4	9	16	25	36	49	64	81	100
$n = 5$ (Pentagonal)	1	5	12	22	35	51	70	92	117	145
$n = 6$ (Hexagonal)	1	6	15	28	45	66	91	120	153	190
$n = 7$ (Heptagonal)	1	7	18	34	55	81	112	148	189	235
$n = 8$ (Octagonal)	1	8	21	40	65	96	133	176	225	280

since the numbers in each column after the first increase linearly, and quadratic in k, since the differences of adjacent numbers in the rows increase linearly. To find the formula, we "triangulate" the kth n-gonal number, partitioning it into a collection of triangular numbers. We illustrate with the 5th hexagonal number 45 in Figure 1.4.

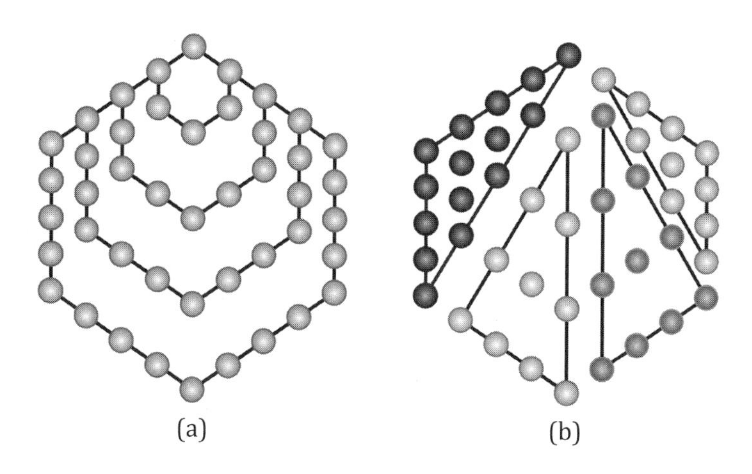

FIGURE 1.4

Hence the 5th hexagonal number 45 equals $T_5 + 3T_4$. In general, the kth n-gonal number equals $T_k + (n-3)T_{k-1}$, since it can be partitioned into one copy of T_k and $n-3$ copies of T_{k-1}, one for each of the $n-3$ diagonals emanating from each vertex of the n-gon. Other expressions for the kth n-gonal number include $k+(n-2)\,T_{k-1}$ and $k^2+(n-4)T_{k-1}$. Each of these formulas is equivalent to $\frac{1}{2}[(n-2)\,k^2-(n-4)\,k]$ for $k \geq 1$ and $n \geq 3$.

Example 1.1. *A test for n-gonal numbers.* Is 903 a triangular number? Is 783 a heptagonal number? A test can be constructed from the formula at the end of the preceding paragraph. Set the candidate x equal to $\frac{1}{2}[(n-2)\,k^2-(n-4)\,k]$ and complete the square on k. To do so we first multiply x by $8(n-2)$:

$$8\,(n-2)\,x = 4[(n-2)^2k^2-(n-4)\,(n-2)k];$$

and then add $(n-4)^2$:

$$\begin{aligned} 8\,(n-2)\,x+(n-4)^2 &= 4\left[(n-2)^2k^2-(n-4)\,(n-2)\,k\right]+(n-4)^2 \\ &= [2\,(n-2)\,k-(n-4)]^2. \end{aligned}$$

So if x is an n-gonal number, then $8(n-2)x + (n-4)^2$ is a perfect square. To find its rank k, we add $n-4$ to the root $2(n-2)k-(n-4)$ and divide by $2(n-2)$.

Is $x = 903$ triangular? Here $n = 3$ so we consider $8x+1 = 8(903)+1 = 7225 = 85^2$, so 903 is a triangular number. Which one? We add -1 to 85 and divide by 2: $k = (85-1)/2 = 42$, hence $903 = T_{42}$.

Is $x = 783$ heptagonal? Here $n = 7$ so we consider $40x + 9 = 40(783) + 9 = 31329 = 177^2$, so 783 is a heptagonal number. Which one? We add 3 to 177 and divide by 10: $k = (177+3)/10 = 18$, hence 783 is the 18th heptagonal number. □

Example 1.2. Many integers appear more than once in Table 1.1. For example, 36 is both triangular and square, every hexagonal number is triangular, and extending the table shows that 210 is both triangular and pentagonal, and 225 is both square and octagonal. Are these coincidences or patterns? We shall answer these questions in later sections in this chapter and in subsequent chapters. □

Polygonal numbers in number theory

Polygonal numbers have long played a role in the theory of numbers. Their properties were investigated by Nicomachus of Gerasa (circa 60–120 CE) and Diophantus of Alexandria (circa 200–284 CE). Pierre de Fermat (1601–1665) wrote that every positive integer is the sum of three or fewer triangular numbers, four or fewer squares, and, in general, n or fewer n-gonal numbers. His proof (if it existed) has never been found. Carl Friedrich Gauss proved the triangular case and on July 10, 1796 wrote in his diary "ΕΥΡΗΚΑ! Num $= \Delta + \Delta + \Delta$." The case for squares was proven in 1770 by Joseph Louis Lagrange (1736–1813) and is known as Lagrange's four-squares theorem. In 1813 Augustin-Louis Cauchy (1789–1857) proved the general case of Fermat's claim. France has honored the three Frenchmen on postage stamps—Fermat in 2001, Lagrange in 1958, and Cauchy in 1989.

Pierre de Fermat, Joseph Louis Lagrange, and Augustin-Louis Cauchy

1.2. Triangular number identities

There are a great many identities relating the triangular numbers. All can be established with elementary algebra, but many can be illustrated visually. Here are several examples.

Example 1.3. *Consecutive sums of consecutive integers.* Consider the sequence of positive integers:

$$1 \;\; 2 \;\; 3 \;\; 4 \;\; 5 \;\; 6 \;\; 7 \;\; 8 \;\; 9 \;\; 10 \;\; 11 \;\; 12 \;\; 13 \;\; 14 \;\; 15 \cdots.$$

If one has an unlimited supply of + and = signs and commas, it is easy to turn the sequence into a sequence of true arithmetic statements:

$$1+2=3, \quad 4+5+6=7+8, \quad 9+10+11+12=13+14+15, \ldots.$$

The above sums—3, 15, and 42—are multiples of the first three triangular numbers, i.e., $3T_1$, $5T_2$, and $7T_3$. In fact, the identity is

$$\begin{aligned} n^2+\left(n^2+1\right)+\cdots+\left(n^2+n\right) &= \left(n^2+n+1\right)+\cdots+\left(n^2+2n\right) \\ &= (2n+1)T_n. \end{aligned}$$

Figure 1.5 gives an illustration for $n = 4$ by considering two arrangements of nine T_4's in the form of collections of unit squares. The numbers along the right side of the figure enumerate the number of unit squares in that row. □

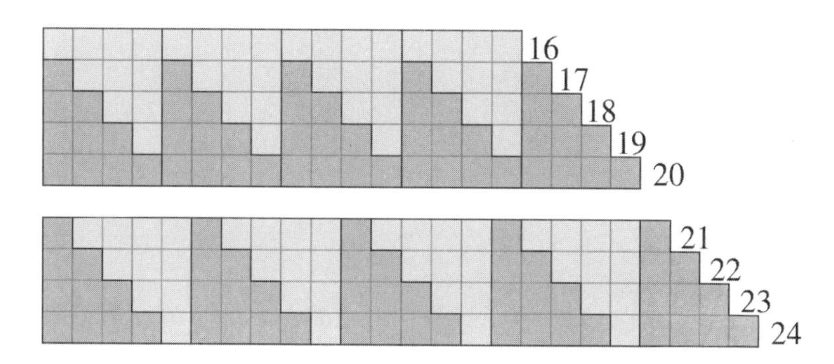

FIGURE 1.5

Example 1.4. *Summing integer cubes.* Sums of integer cubes are related to triangular numbers in the formula

$$1^3+2^3+\cdots+n^3=(1+2+\cdots+n)^2=T_n^2$$

often encountered in precalculus mathematics. Figure 1.1(a) illustrates $k = T_k - T_{k-1}$, and Figure 1.3(b) illustrates $k^2 = T_k + T_{k-1}$. Multiplying these together yields

$$k^3 = k \cdot k^2 = \left(T_k - T_{k-1}\right)\left(T_k + T_{k-1}\right) = T_k^2 - T_{k-1}^2. \tag{1.1}$$

Summing (1.1) and observing that the sum telescopes yields the formula for the sum of the first n positive integer cubes (recall that $T_0 = 0$):

$$\sum_{k=1}^{n} k^3 = \sum_{k=1}^{n} (T_k^2 - T_{k-1}^2) = T_n^2 - T_0^2 = T_n^2.$$

In Figure 1.6 [Golomb, 1965] we see an illustration of this result, where we represent k^3 by k copies of squares with area k^2 from 1 to n (shown here for $n = 6$). When k is even, two squares overlap, but the area of the overlap is the same as the area of the square (in white) not covered by the shaded squares. □

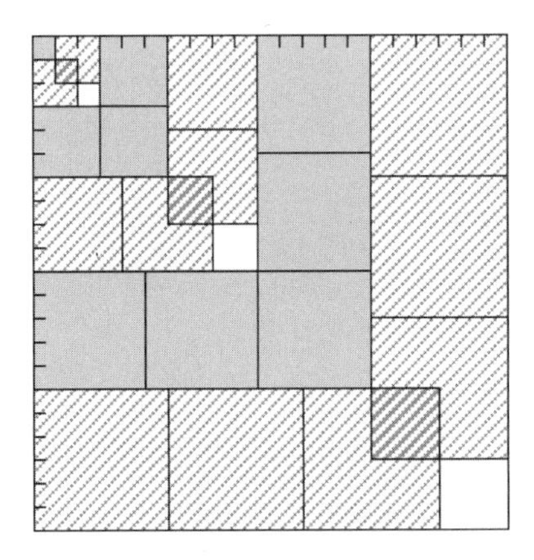

FIGURE 1.6

Example 1.5. *Summing squares of consecutive triangular numbers.* While the difference of the squares of two consecutive triangular numbers is a cube, the sum of the squares of two consecutive triangular numbers is triangular, $T_k^2 + T_{k-1}^2 = T_{k^2}$, as illustrated in Figure 1.7 for $k = 4$. We represent T_4^2 with T_4 copies of T_4 and similarly for T_3^2. □

Example 1.6. *Sums of consecutive triangular numbers.* Analogous to Example 1.3 we consider the sequence of triangular numbers:

$$1\ 3\ 6\ 10\ 15\ 21\ 28\ 36\ 45\ 55\ 66\ 78\ 91\ 105\ 120\ 136\ 153\ 171 \cdots.$$

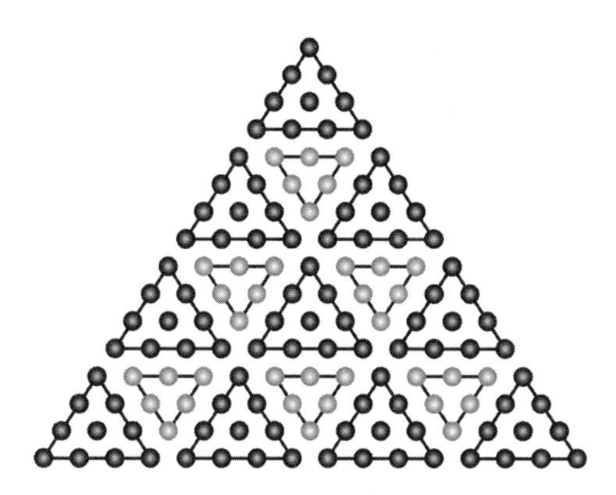

FIGURE 1.7

If, as in Example 1.3, we have an unlimited supply of + and = signs and commas, we can also turn this sequence into a sequence of true arithmetic statements:

$$1 + 3 + 6 = 10,$$
$$15 + 21 + 28 + 36 = 45 + 55,$$
$$66 + 78 + 91 + 105 + 120 = 136 + 153 + 171, \ldots.$$

The pattern is

$$T_{n^2-n-1} + T_{n^2-n} + \cdots + T_{n^2-1} = T_{n^2} + T_{n^2+1} + \cdots + T_{n^2+n-2}.$$

For $n = 4$ the pattern is $T_{11} + T_{12} + T_{13} + T_{14} + T_{15} = T_{16} + T_{17} + T_{18}$, which we now illustrate [Nelsen and Unal, 2012]. Figure 1.8 shows that

$$T_{16} + T_{17} + T_{18} = T_{15} + T_{14} + T_{13} + 1 \cdot 4^2 + 3 \cdot 4^2 + 5 \cdot 4^2.$$

But $1 \cdot 4^2 + 3 \cdot 4^2 + 5 \cdot 4^2 = 9 \cdot 4^2 = 12^2 = T_{11} + T_{12}$, as illustrated in Figure 1.9, which completes the illustration. □

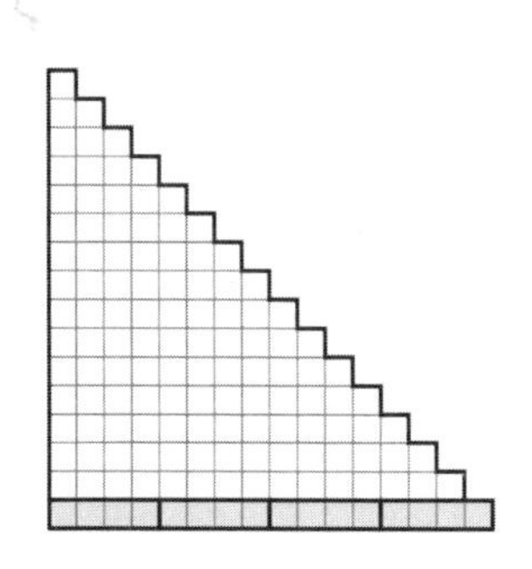

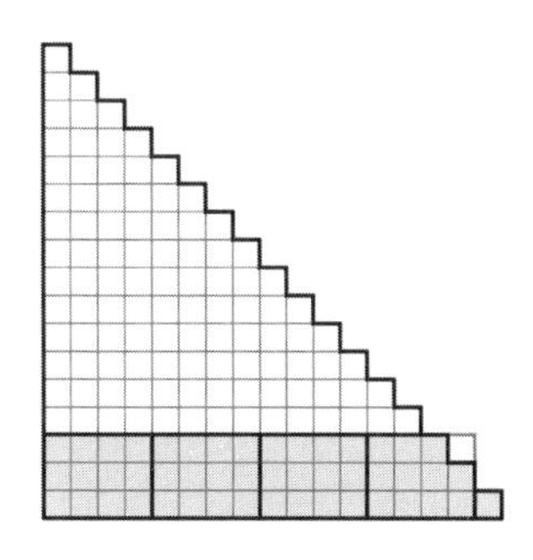

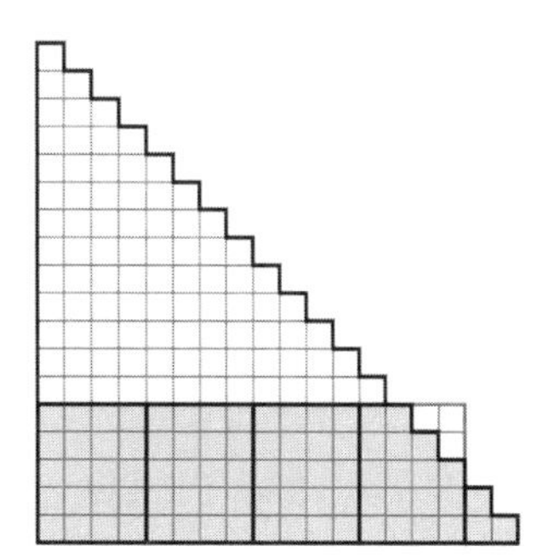

FIGURE 1.8

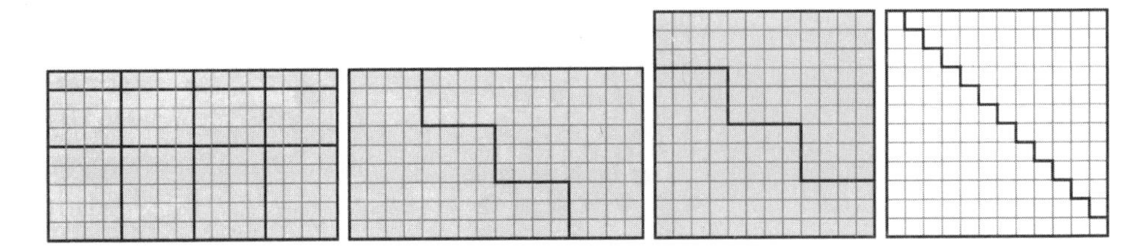

FIGURE 1.9

In the following theorem we relate the factors of one triangular number to a pair of larger triangular numbers with a triangular sum. The theorem is surprisingly useful, as the two examples following the proof illustrate.

Theorem 1.1. *Let n, p, and q be positive integers. Then*

$$T_n = pq \text{ if and only if } T_{n+p} + T_{n+q} = T_{n+p+q}. \tag{1.2}$$

Proof. In Figure 1.10 we illustrate T_{n+p+q} for $n = 6$, $p = 3$, and $q = 7$. Counting the dots using the inclusion-exclusion principle yields $T_{n+p+q} = T_{n+p} + T_{n+q} - T_n + pq$, from which (1.2) follows. The theorem can also be established algebraically. □

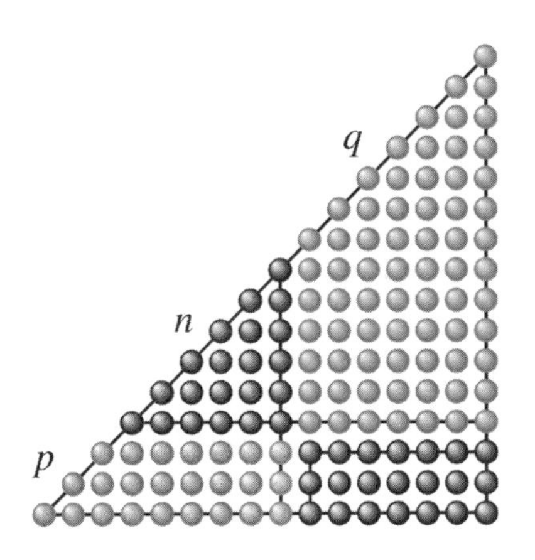

FIGURE 1.10

Figure 1.10 shows that $T_6 = 21 = 3 \cdot 7$ is equivalent to $T_9 + T_{13} = T_{16}$. Similar but perhaps more interesting sums are encountered in the following examples.

Example 1.7. *Triangular sums and Pythagorean triples.* Let k be a positive integer. Then (1.1) yields the following triangular sums of two triangular numbers. Let k be a positive integer; then:

(a) with $T_{2k} = k(2k+1)$, we have
$$T_{3k} + T_{4k+1} = T_{5k+1};$$
(b) with $T_{4k} = k(8k+2)$, we have
$$T_{5k} + T_{12k+2} = T_{13k+2};$$
(c) with $T_{6k+2} = (2k+1)(9k+3)$, we have
$$T_{8k+3} + T_{15k+5} = T_{17k+6};$$
(d) with $T_{12k+3} = (8k+2)(9k+3)$, we have
$$T_{20k+5} + T_{21k+6} = T_{29k+8}.$$

Each triple of coefficients of k in the ranks of the triangular numbers in the sums—$(3,4,5)$, $(5,12,13)$, $(8,15,17)$, and $(20,21,29)$—is a *Pythagorean triple*, i.e., a triple (a,b,c) where $a^2 + b^2 = c^2$. Observe some patterns in the factors of the given triangular number T_n and the triples:

i) If the coefficient of k in one factor of T_n is 1, then the even leg and hypotenuse of the triple are consecutive integers.
ii) If the coefficient of k in one factor of T_n is 2, then the odd leg and hypotenuse of the triple differ by 2.
iii) If the coefficients of k in the factors of T_n are consecutive integers, then the legs of the triple are consecutive integers.

Pythagorean triples are the subject of Chapter 4. □

Example 1.8. *Square triangular and oblong triangular numbers.* Some numbers, such as 1 and 36, are both square and triangular; and others, such as 6 and 210, are both oblong and triangular. Are there others? Setting $q = p$ in (1.2) yields
$$T_n = p^2 \text{ if and only if } T_{n+2p} = 2T_{n+p} = (n+p)(n+p+1),$$
that is, each square triangular number corresponds to a larger oblong triangular number. Setting $q = p+1$ in (1.2) yields
$$T_n = p(p+1) \text{ if and only if } T_{n+2p+1} = T_{n+p} + T_{n+p+1} = (n+p+1)^2,$$
that is, each oblong triangular number corresponds to a larger square triangular number. Combining the two equivalences in two ways yields
$$T_n = p^2 \text{ if and only if } T_{3n+4p+1} = (2n+3p+1)^2 \tag{1.3}$$
and
$$T_n = p(p+1) \text{ if and only if } T_{3n+4p+3} = (2n+3p+2)(2n+3p+3). \tag{1.4}$$

Equivalence (1.3) generates infinitely many square triangular numbers $T_1 = 1^2, T_8 = 6^2, T_{49} = 35^2, T_{288} = 204^2$, etc., and (1.4) generates infinitely many oblong triangular numbers $T_3 = 2 \cdot 3, T_{20} = 14 \cdot 15, T_{119} = 84 \cdot 85, T_{696} = 492 \cdot 493$, etc. We shall continue our study of *multi-polygonal numbers* (numbers polygonal in two or more ways) in the next two chapters. □

You may have noticed that the nth triangular number is a binomial coefficient, i.e., $T_n = \binom{n+1}{2}$, as illustrated with a portion of Pascal's triangle in Figure 1.11(a). One explanation for this is that each is equal to $n(n+1)/2$, but this answer sheds little light on *why* it is true. Here is a better explanation.

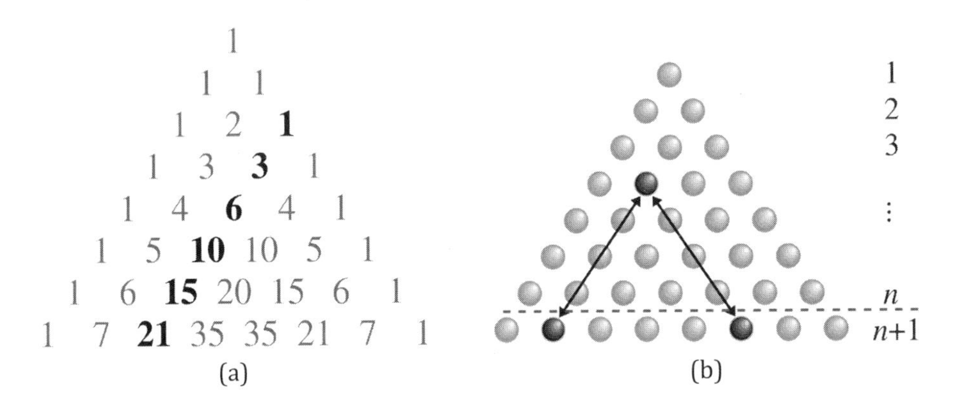

FIGURE 1.11

Theorem 1.2. *There exists a one-to-one correspondence between a set of T_n objects and the set of two-elements subsets of a set with $n + 1$ objects.*

Proof. See Figure 1.11(b) [Larson, 1985], and recall that the binomial coefficient $\binom{k}{2}$ is the number of ways to choose 2 elements from a set of k objects. The arrows denote the correspondence between an element of a set with T_n objects and a pair of elements from a set with $n + 1$ objects. □

We conclude this section with a recurrence relation for the triangular numbers that generalizes to a recurrence relation for all the n-gonal numbers [Edgar, 2017]. We begin with a general inclusion-exclusion principle for triangular numbers.

Theorem 1.3. *If $0 \leq a, b, c \leq n$ and $a + b + c \geq 2n$, then*

$$T_n = T_a + T_b + T_c - T_{a+b-n} - T_{b+c-n} - T_{c+a-n} + T_{a+b+c-2n}. \tag{1.5}$$

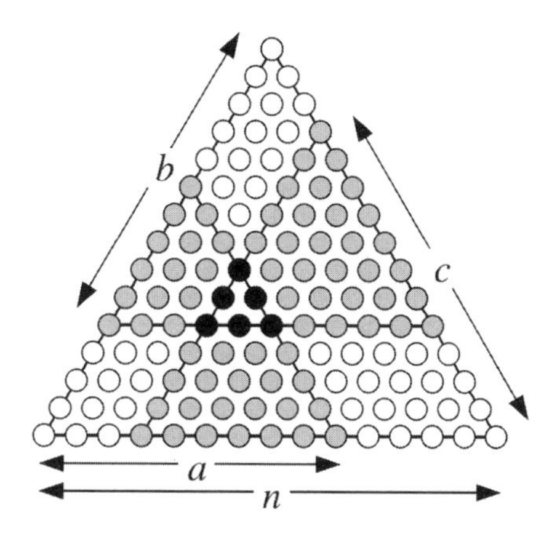

FIGURE 1.12

Proof. See Figure 1.12 and compute T_n using the inclusion-exclusion principle. □

Corollary 1.4. *For all $k \geq 2$, $T_{k+1} = 3T_k - 3T_{k-1} + T_{k-2}$.*

Proof. Set $(n, a, b, c) = (k+1, k, k, k)$ in (1.5), noting that $a + b + c = 3k \geq 2(k+1) = 2n$ since $k \geq 2$. □

With initial conditions $T_0 = 0$, $T_1 = 1$, and $T_2 = 3$ the identity in Corollary 1.4 produces all the triangular numbers. Now let p_k^n denote the kth n-gonal number. In the preceding section we saw that every n-gonal number is a sum of triangular numbers, i.e., $p_k^n = T_k + (n-3)T_{k-1}$, which yields the following corollary.

Corollary 1.5. *For all $k \geq 2$ and $n \geq 3$, $p_{k+1}^n = 3p_k^n - 3p_{k-1}^n + p_{k-2}^n$.*

With initial conditions $p_0^n = 0$, $p_1^n = 1$, and $p_2^n = n$ this identity generates all the n-gonal numbers.

Example 1.9. To compute the heptagonal numbers p_k^7 we use the identity in the corollary with $p_0^7 = 0$, $p_1^7 = 1$, and $p_2^7 = 7$:

$$p_3^7 = 3 \cdot 7 - 3 \cdot 1 + 0 = 18,$$
$$p_4^7 = 3 \cdot 18 - 3 \cdot 7 + 1 = 34,$$
$$p_5^7 = 3 \cdot 34 - 3 \cdot 18 + 7 = 55, \text{ etc.}$$
□

1.3. Oblong numbers and the infinitude of primes

The earliest proof that there are infinitely many prime numbers is probably Euclid's in the *Elements* (Book IX, Proposition 20). After more than 2000 years, it is difficult to find a better one. Here is a version of Euclid's proof employing the dimensions n and $n+1$ of the nth oblong number O_n, created by Ernst Eduard Kummer (1810–1893) in 1873 [Ribenboim, 1996]. In the proof we show that the dimensions n and $n+1$ of O_n have no common prime factors. Kummer's proof is indirect like virtually all proofs of the infinitude of primes, since one definition of "infinitely many" is "*not* finitely many."

Theorem 1.6. *There are infinitely many prime numbers.*

Proof. Assume there are only k primes, $p_1 = 2, p_2 = 3, p_3 = 5, \ldots, p_k$. Let $N = p_1 p_2 p_3 \cdots p_k$. Since $N+1$ is larger than p_k it is not prime, and thus $N+1$ has a prime divisor p_j in common with N. Since p_j divides both N and $N+1$, it divides $(N+1)-N = 1$, which is impossible. Hence there are infinitely many primes. □

Euclid primes

Numbers of the form $N_k = p_1 p_2 p_3 \cdots p_k$ are called *primorials* (from *prime* and *factorial*), and the number $E_k = N_k + 1$ is called a *Euclid number*. The first five Euclid numbers 3, 7, 31, 211, 2311 are prime (and called *Euclid primes*), however $E_6 = 30031 = 59 \cdot 509$. It is not known if the number of Euclid primes is finite or infinite.

1.4. Pentagonal and other figurate numbers

The kth pentagonal number P_k can be computed by $T_k + 2T_{k-1}$ and by $k^2 + T_{k-1}$. There are two other less obvious but useful ways, illustrated in Figure 1.13.

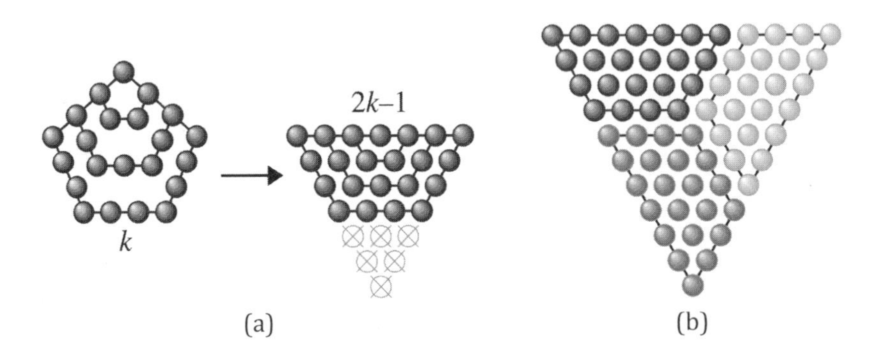

FIGURE 1.13

Distorting the pentagonal shape of P_k into the trapezoid in Figure 1.13(a) shows that $P_k = T_{2k-1} - T_{k-1}$, and Figure 1.13(b) shows how three copies of the trapezoid yield a triangle, so that $T_{3k-1} = 3P_k$ and hence $P_k = T_{3k-1}/3$ (in the next chapter we show that T_{3k-1} is always divisible by 3). We shall use this result in Chapter 3 to find infinitely many triangular pentagonal numbers.

Examination of Table 1.1 shows that every hexagonal number H_k is a triangular number with an odd rank, and that their common value is the product of their ranks, i.e., $H_k = T_{2k-1} = k(2k-1)$. See Figure 1.14 for an illustration when $k = 5$, i.e., $H_5 = T_9 = 5 \cdot 9$, on the cover of the January 2014 issue of *The College Mathematics Journal*.

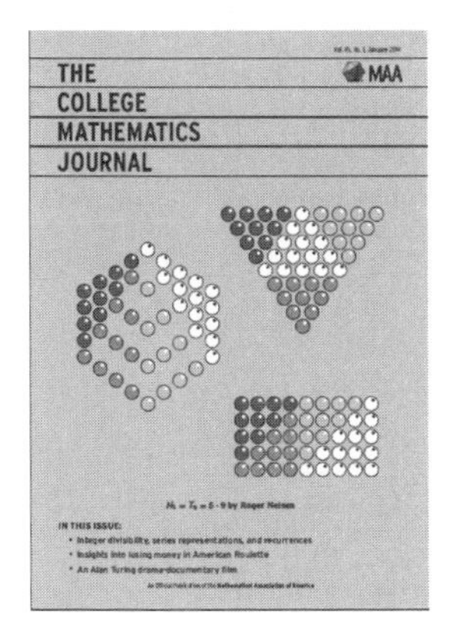

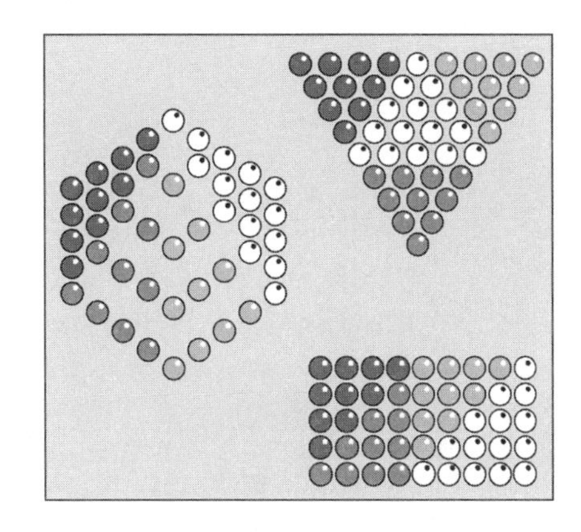

FIGURE 1.14. (This illustration originally appeared on the cover of the January 2014 issue of *The College Mathematics Journal*.)

Analogous to our expression of a pentagonal number as the difference of two triangular numbers, each octagonal number is the difference of two squares. In Figure 1.15 we illustrate $\mathrm{Oct}_k = k(3k-2) = (2k-1)^2 - (k-1)^2$ for $k = 4$.

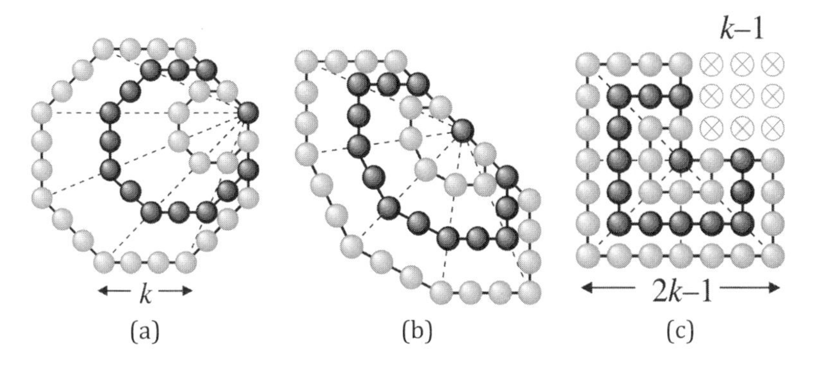

FIGURE 1.15

Chinese checkers and star numbers

Chinese checkers is a board game usually played with marbles on a wooden board with holes arranged in a star-like pattern, as illustrated below on the left. The standard board has $S_5 = 121$ holes, where S_n is the nth *star number*. For $n \geq 2$, $S_n = 12T_{n-1} + 1 = 6n(n-1) + 1$, seen in the middle image below. The second star number $S_2 = 13$ appears in the Great Seal of the United States on the green side of the one-dollar bill, as shown below on the right

Chinese checkers and the star numbers S_5 and S_2

In spite of its name, Chinese checkers did not originate in China and is unrelated to ordinary checkers. The game was invented in Germany in the 1890s and first produced in the U.S. in 1928. The name "Chinese checkers" was chosen for marketing purposes.

1.5. Polite numbers

An integer is *polite* if it can be written as the sum of two or more consecutive positive integers in at least one way. For example, every triangular number (greater than 1) is polite, and every pentagonal number (greater than 1) is polite (see Figure 1.5). Some integers have more than one polite representation, for example, 15 has three polite representations: $7+8$, $4+5+6$, and $1+2+3+4+5$. Numbers that can be written as the sum of two or more consecutive positive integers greater than 1 are called *trapezoidal*, e.g., all pentagonal numbers greater than 1 are trapezoidal, and 15 is both triangular and trapezoidal. So polite numbers are triangular or trapezoidal (or both). So we ask—which integers are polite?

Theorem 1.7. *A positive integer n is polite if and only if it is not a power of 2.*

Proof. Assume n is not a power of 2, i.e., $n = 2^a(2b+1)$, where $a \geq 0$ and $b \geq 1$, and let $m = \min\{2^{a+1}, 2b+1\}$ and $M = \max\{2^{a+1}, 2b+1\}$.

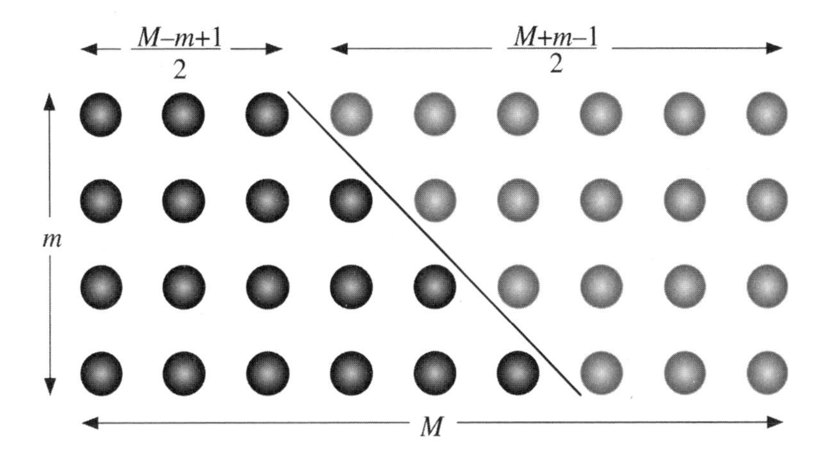

FIGURE 1.16. (This illustration originally appeared in C. L. Frenzen, "Proof without words: Sums of consecutive positive integers", *Mathematics Magazine* **70** (1997), 294.

Thus $2n = mM$, which we represent by $2n$ balls in an $m \times M$ array, as shown in Figure 1.16 [Frenzen, 1997].

Then $M \geq 3$, $m \geq 2$, $\frac{M+m-1}{2} \geq 2$, $\frac{M-m+1}{2} \geq 1$, and $\frac{M+m-1}{2} - \frac{M-m+1}{2} \geq 1$ so that

$$n = \left(\frac{M-m+1}{2}\right) + \left(\frac{M-m+1}{2} + 1\right) + \cdots + \left(\frac{M+m-1}{2}\right).$$

Now assume that n is polite, that is, $n = (a+1) + (a+2) + \cdots + (a+k)$ with $a \geq 0$ and $k \geq 2$, so that $2n = k(k+2a+1)$. Since k and $k+2a+1$ have opposite parity, one of them is odd and hence n cannot be a power of 2. □

When $2b+1$ is a composite odd number in the above proof, $n = 2^a(2b+1)$ has polite representations different from the one in Figure 1.16. For example, the figure illustrates $18 = 3+4+5+6$. But if we partition a set of 36 balls in a 3×12 array, we have $18 = 5+6+7$. When $2b+1$ is prime, n has only the polite representation in Figure 1.16, e.g., when $n = 14$, the only polite representation is $14 = 2+3+4+5$.

Example 1.10. *Non-trapezoidal polite numbers.* If $n = 2^a(2b+1)$, where $a \geq 0$ and $b \geq 1$, is a non-trapezoidal polite number, then n is a triangular number with $2b+1$ prime. In this case the balls in Figure 1.16 represent an oblong number partitioned into two triangular numbers, as in Figure 1.3(a). There are only two forms for such triangular numbers, $T_{2^k} = 2^{k-1}(2^k + 1)$ with $2^k + 1$ prime, and $T_{2^k-1} = 2^{k-1}(2^k - 1)$ with $2^k - 1$ prime [Jones and Lord, 1999] . When $2^k + 1$ is prime, k is a power of 2 and $2^k + 1$ is a *Fermat prime.* When $2^k - 1$ is prime, so is k (and $2^k - 1$ is known as a *Mersenne prime*) and T_{2^k-1} is an *even perfect number*, the subject of Chapter 7. As a consequence, the non-trapezoidal polite numbers are rather rare, the only ones less than 2,000,000,000 are

$$3, 6, 10, 28, 136, 496, 8128, 32896, \text{and } 33550336. \qquad \square$$

1.6. Three-dimensional figurate numbers

Figure 1.17 shows how cannonballs were stored at Fort Monroe in Hampton Roads, Virginia, a Union military installation in 1861. The piles of cannonballs had both triangular and rectangular bases.

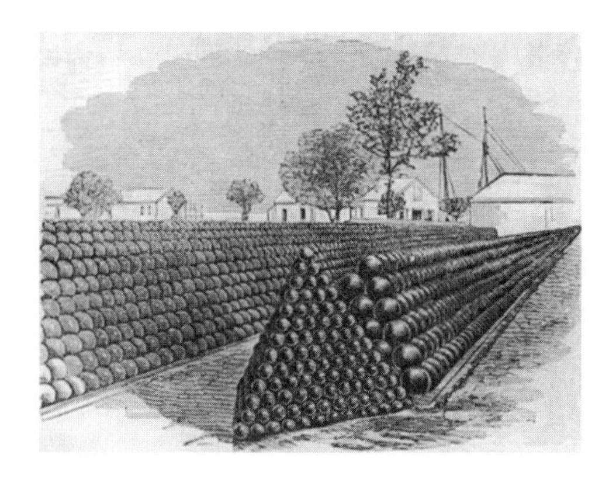

FIGURE 1.17

Example 1.11. *Tetrahedral numbers.* How many cannonballs are there in a tetrahedral pile? In Figure 1.18(a) we see cannonballs in a pile whose shape is that of a regular tetrahedron. The cannonballs in each layer are arranged in a triangle, and the number of balls in each layer is a triangular number. The number of cannonballs in such a pile with n layers is the *tetrahedral number* Tet_n. Figure 1.18(b) shows that $\text{Tet}_5 = 1 + 3 + 6 + 10 + 15 = 35$.

In Figure 1.19 [Zerger, 1990] we see a $T_n \times (n + 2)$ rectangle of unit squares partitioned into three copies of $T_1 + T_2 + \cdots + T_n$, and hence $3(T_1 + T_2 + \cdots + T_n) = T_n(n + 2)$, and thus

$$\text{Tet}_n = \sum_{k=1}^{n} T_k = \frac{n(n+1)(n+2)}{6}.$$

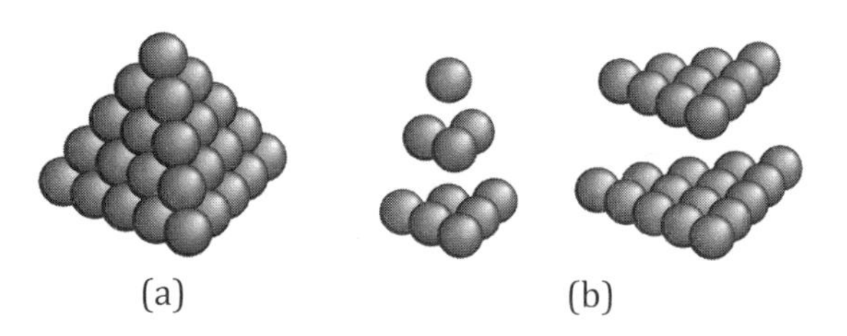

FIGURE 1.18

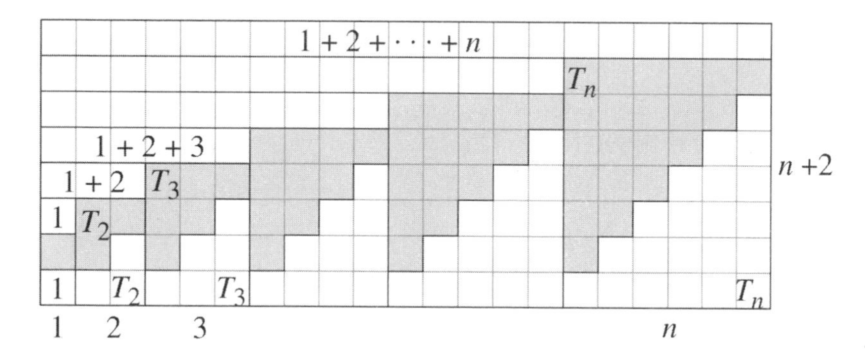

FIGURE 1.19

Note that like T_n, Tet_n is also a binomial coefficient: $\mathrm{Tet}_n = \binom{n+2}{3}$. □

Example 1.12. *Pyramidal numbers.* Analogous to the tetrahedral numbers we have the *pyramidal* (or *square pyramidal*) *numbers*, represented by the number of cannonballs stacked in a pyramidal pile with a square base, as seen in Figure 1.20(a).

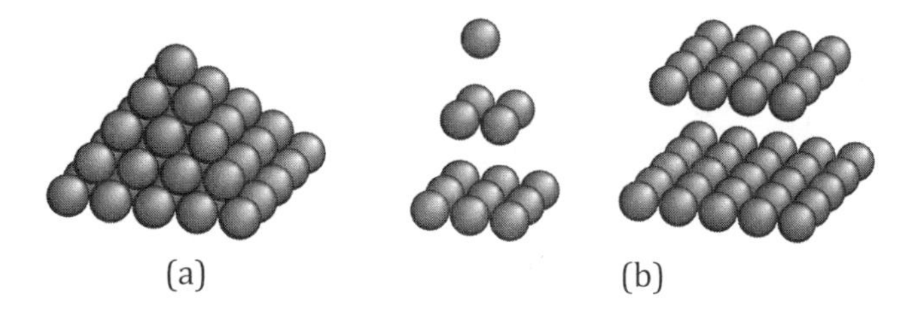

FIGURE 1.20

The number of cannonballs in each layer is a square, and so the pyramidal number Pyr_n is the sum of the first n squares. Figure 1.20(b) shows that $\mathrm{Pyr}_5 = 1 + 4 + 9 + 16 + 25 = 55$.

In Figure 1.21 we see a representation of Pyr_5 with unit cubes rather than cannonballs. The white cubes represent Tet_5 while the gray cubes represent Tet_4, so that $\mathrm{Pyr}_5 = \mathrm{Tet}_4 + \mathrm{Tet}_5$.

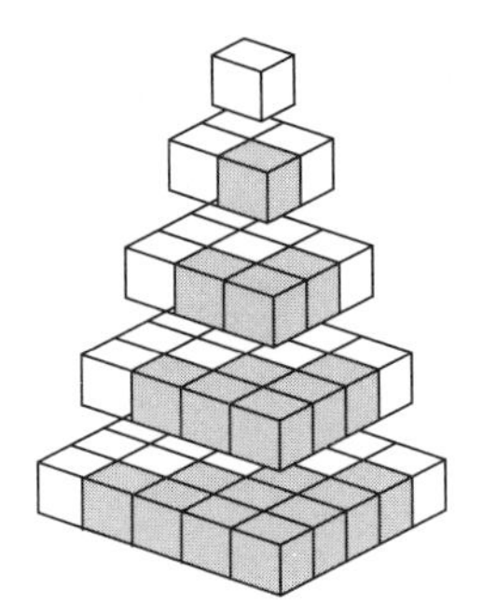

FIGURE 1.21

Hence in general we have

$$\begin{aligned}\mathrm{Pyr}_n &= \sum_{k=1}^{n} k^2\\ &= \mathrm{Tet}_{n-1} + \mathrm{Tet}_n\\ &= \frac{(n-1)n(n+1)}{6} + \frac{n(n+1)(n+2)}{6}\\ &= \frac{n(n+1)(2n+1)}{6}.\end{aligned}$$

For another derivation of the formula for Pyr_n as a sum of squares, see Figure 1.22 [Gardner, 1973; Kalman, 1991] . Three sets of the first n squares (for $n = 5$) are shown in Figure 1.22(a).

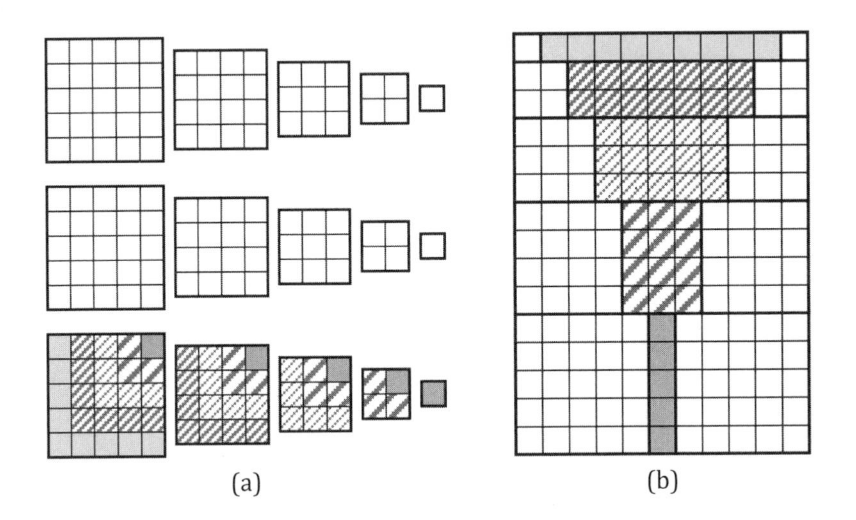

FIGURE 1.22

If we slice up the squares in the third set as indicated by the different shadings, all the unit squares can be rearranged into the rectangle in Figure 1.22(b) with base $n+1+n = 2n+1$ and height $1+2+\cdots+n = T_n$, so that

$$3\left(1^2 + 2^2 + \cdots + n^2\right) = (2n+1)\,T_n = \frac{n(n+1)(2n+1)}{2},$$

or $3\,\mathrm{Pyr}_n = (2n+1)\,T_n$, from which the formula for Pyr_n follows upon division by 3. □

Example 1.13. *Sums of pentagonal numbers.* For our final example of three-dimensional figurate numbers, we sum pentagonal numbers. In Figure 1.23(a) we illustrate $P_k = k^2 + T_{k-1}$ with a vertical arrangement of k^2 unit cubes joined to a horizontal arrangement of T_{k-1} unit cubes for $k = 1, 2, 3$, and 4 $(= n)$. The resulting configurations of cubes can now be nested to form a pile with base area T_n and height n as shown in Figure 1.23(b) [Miller, 1993] so that

$$\sum_{k=1}^{n} P_k = nT_n = \frac{n^2(n+1)}{2}.$$ □

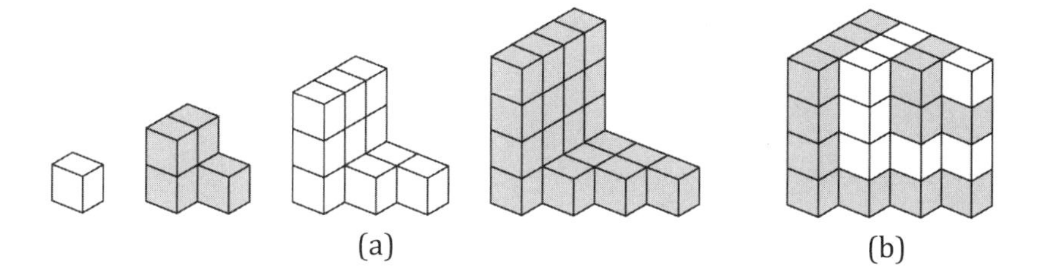

FIGURE 1.23

For a comprehensive overview of figurate numbers and their properties, see [Deza and Deza, 2012].

1.7. Exercises

1.1 In Figure 1.3 we saw illustrations of two relationships between triangular, oblong, and square numbers. Two more relationships are $k^2 + O_k = T_{2k}$ and $(k+1)^2 + O_k = T_{2k+1}$. Can you illustrate these with figures similar to Figure 1.3?

1.2 Show that the sum of the first n $(n \geq 0)$ powers of 9 is a triangular number, e.g., $1 + 9 = T_4$, $1 + 9 + 81 = T_{13}$, etc.

1.3 One triangular number can be twice another, as in $T_3 = 2T_2$. Are there other instances of this phenomenon?

1.4 In Section 1.4 we saw that each pentagonal number is the difference of two triangular numbers. Prove that the same is true of odd squares, e.g., $5^2 = T_7 - T_2$, $7^2 = T_{10} - T_3$, etc. Can you illustrate the result with a picture?

1.5 In Table 1.1 it appears that the only prime n-gonal numbers are the 2nd p-gonal numbers for p a prime. Prove that this is indeed the case.

1.6 *Centered triangular numbers.* The centered triangular number c_n enumerates the number of dots in an array with one central dot surrounded by dots in triangular borders. The number $c_5 = 46$ is illustrated in Figure 1.24(a).

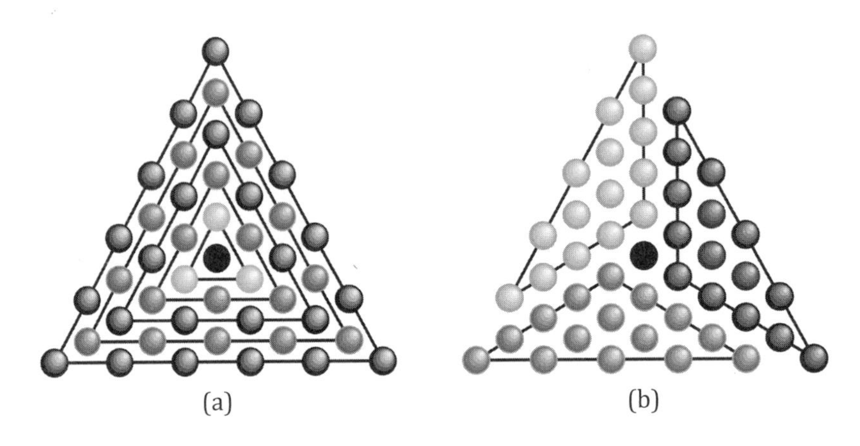

(a) (b)

FIGURE 1.24

In Figure 1.24(b) we see that $c_n = 1 + 3T_n$. Show that another formula for c_n is $c_n = T_{n-1} + T_n + T_{n+1}$ for $n \geq 2$.

1.7 Show that the sum of the first n odd squares is a tetrahedral number. [Hint: See Figure 1.25 where we represent integers by unit cubes.]

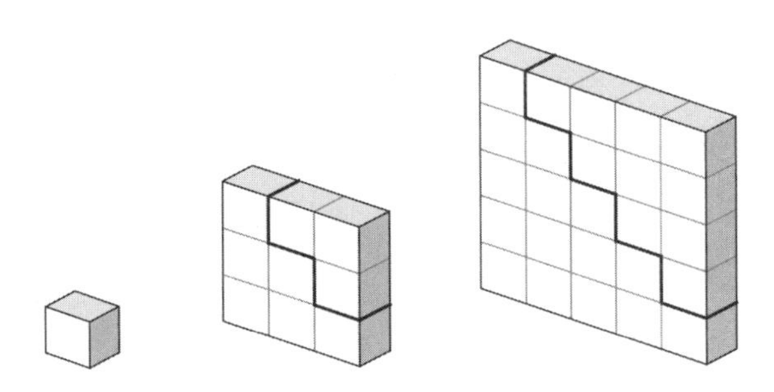

FIGURE 1.25

1.8 The sequence $\{\text{Tet}_n\}_{n=1}^{\infty}$ of tetrahedral numbers is $\{1, 4, 10, 20, 35, 56, 84, 120, \ldots\}$, while the sequence $\{\text{Pyr}_n\}_{n=1}^{\infty}$ of pyramidal numbers is $\{1, 5, 14, 30, 55, 91, 140, 204, \ldots\}$. Observe that the sequence $\{4\,\text{Pyr}_n\}_{n=1}^{\infty}$ is $\{4, 20, 56, 120, \ldots\}$. Formulate a theorem and prove it. Can you illustrate the theorem by stacking and re-stacking cannonballs?

1.9 The *octahedral numbers* are defined as the number of cannonballs in octahedral piles, as illustrated in Figure 1.26 for the first five octahedral numbers 1, 6, 19, 44, and 85.

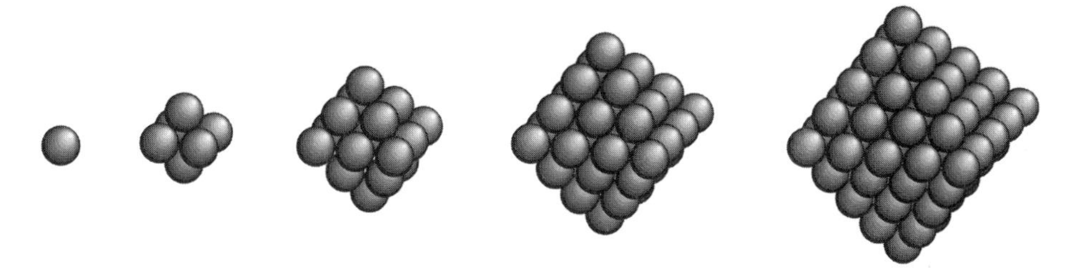

FIGURE 1.26

Find a formula for the nth octahedral number.

CHAPTER 2

Congruence

Mathematics is the queen of the sciences and number theory is the queen of mathematics.
Johann Carl Friedrich Gauss

No one who lacks an acquaintance with congruences can claim to know much about number theory.
Underwood Dudley

One of the most important relations in number theory—and one of the most beautiful—is congruence. We say that *a is congruent to b modulo m*, and write $a \equiv b \pmod{m}$, if and only if m divides $a - b$ where m, called the *modulus*, is a positive integer. Equivalently, $a \equiv b \pmod{m}$ if and only if $a = b + km$ for some integer k. The congruence symbol $\equiv$ was first employed by Gauss about 1800. For example, congruences modulo 2 express whether an integer is even (congruent to 0 modulo 2) or odd (congruent to 1 modulo 2). When a number k divides a number n one often writes $k|n$, i.e., $a \equiv b \pmod{m}$ is equivalent to $m|(a - b)$.

2.1. Congruence results for triangular numbers

In this section we consider some congruence theorems for triangular numbers that will be useful in later chapters, for example, when we consider Pythagorean triples in Chapter 4 and perfect numbers in Chapter 7. Here we evaluate T_n modulo m for m equal to 2, 3, and 5. In each instance we partition triangular numbers to illustrate identities that establish the congruences.

In the $n = 3$ row of Table 1.1 the first two triangular numbers are odd, the next two are even, the next two odd, etc., which leads to the following theorem.

Theorem 2.1. *For $n \geq 1$ we have*

$$T_n \equiv \begin{cases} 0 \pmod{2}, & n \equiv 0 \text{ or } 3 \pmod{4}, \\ 1 \pmod{2}, & n \equiv 1 \text{ or } 2 \pmod{4}. \end{cases}$$

Proof. In Figure 2.1 we illustrate that when $k = 6$, $T_{4k} = 10T_k + 6T_{k-1}$, $T_{4k+1} = 14T_k + 2T_{k-1} + 1$, $T_{4k+2} = 14T_k + 2T_{k+1} + 1$, and $T_{4k+3} = 10T_k + 6T_{k+1}$. These four equations can now be verified by simple algebra using $T_n = n(n+1)/2$. Hence we have $T_{4k} \equiv T_{4k+3} \equiv 0 \pmod{2}$ and $T_{4k+1} \equiv T_{4k+2} \equiv 1 \pmod{2}$. □

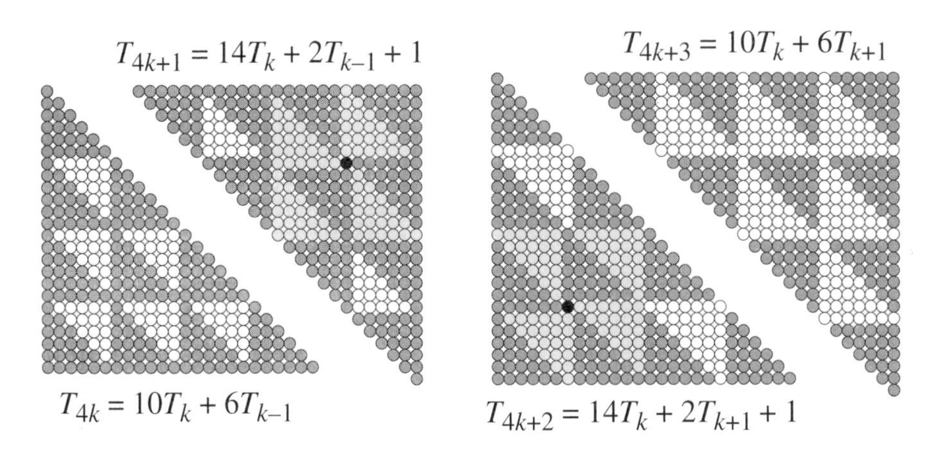

FIGURE 2.1

Theorem 2.2. *For $n \geq 1$ we have*

$$T_n \equiv \begin{cases} 1 \pmod{3}, & n \equiv 1 \pmod{3}, \\ 0 \pmod{3}, & n \not\equiv 1 \pmod{3}. \end{cases}$$

Proof 1. In Figure 2.2 we show that $T_{3k-1} = 6T_{k-1} + 3T_k$, $T_{3k} = 6T_k + 3T_{k-1}$, and $T_{3k+1} = 9T_k + 1$ for $k = 7$, and the general case can be verified by algebra. Hence $T_{3k-1} \equiv T_{3k} \equiv 0 \pmod{3}$ and $T_{3k+1} \equiv 1 \pmod{3}$. □

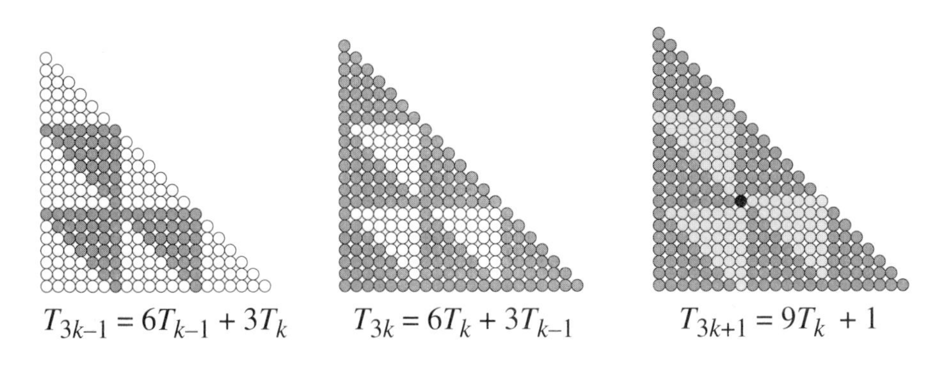

FIGURE 2.2

Observe that, as a bonus, we have $T_{3k+1} \equiv 1 \pmod 9$. This congruence will be useful in Chapter 7 when we consider perfect numbers.

Proof 2. Corollary 1.4 states that for all $k \geq 2$, $T_{k+1} = 3T_k - 3T_{k-1} + T_{k-2}$, so that $T_{k+1} \equiv T_{k-2} \pmod 3$. Hence any two triangular numbers whose ranks differ by a multiple of 3 are congruent modulo 3, and the theorem follows from $T_1 \equiv 1 \pmod 3$, $T_2 \equiv T_3 \equiv 0 \pmod 3$. □

Theorem 2.3. *For $n \geq 1$ we have*

$$T_n = \begin{cases} 0 \pmod 5, & n \equiv 0 \text{ or } 4 \pmod 5, \\ 1 \pmod 5, & n \equiv 1 \text{ or } 3 \pmod 5, \\ 3 \pmod 5, & n \equiv 2 \pmod 5. \end{cases}$$

Proof. In Figure 2.3 we show that $T_{5k} = 15T_k + 10T_{k-1}$, $T_{5k+1} = 20T_k + 5T_{k-1} + 1$, $T_{5k+2} = 25T_k + 3$, $T_{5k+3} = 20T_k + 5T_{k+1} + 1$, and $T_{5k+4} = 15T_k + 10T_{k+1}$ for $k = 5$, and again the general case can be verified by algebra. Hence we have $T_{5k} \equiv T_{5k+4} \equiv 0 \pmod 5$, $T_{5k+1} \equiv T_{5k+3} \equiv 1 \pmod 5$, and $T_{5k+2} \equiv 3 \pmod 5$. □

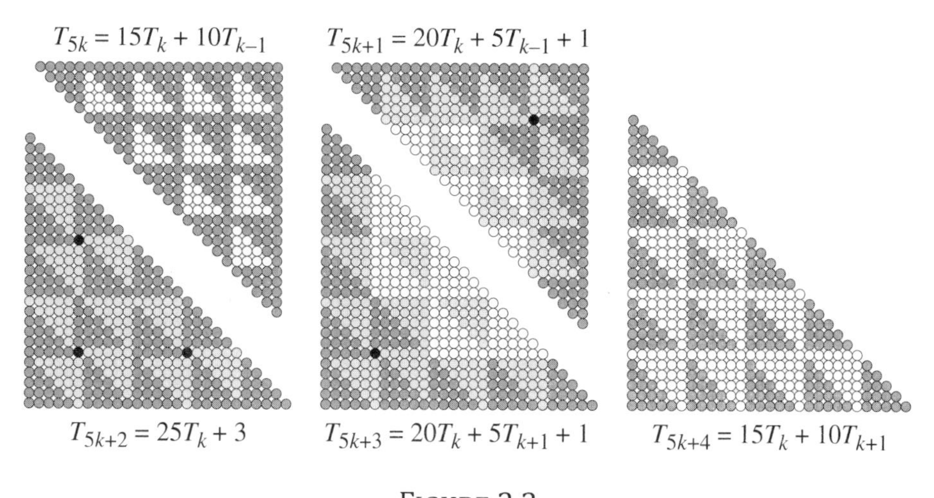

FIGURE 2.3

2.2. Congruence results for other figurate numbers

In Example 1.1 we presented a test for triangular numbers: 8 times a triangular number plus 1 is a square. Now we present a visual illustration of that relationship and express it as a congruence result for odd squares.

Lemma 2.4. *For $n \geq 1$ we have $(2n+1)^2 = 8T_n + 1$, so that $(2n+1)^2 \equiv 1 \pmod 8$.*

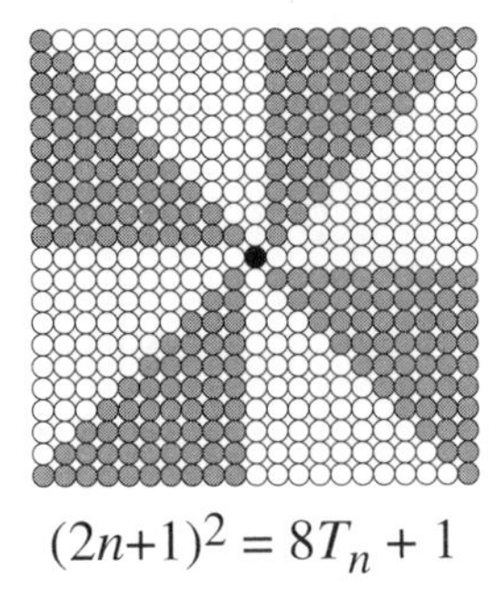

$(2n+1)^2 = 8T_n + 1$

FIGURE 2.4

Proof. See Figure 2.4 for $n = 10$. Algebra verifies the result in general. □

In Example 3.2 in the next chapter we evaluate squares modulo 3 using triangles, similar to our evaluation of triangular numbers modulo 3 in Theorem 2.2.

Lemma 2.4 enables us to prove the following two theorems.

Theorem 2.5. *If the positive integer m is not divisible by 2 or 3, then $m^2 \equiv 1 \pmod{24}$.*

Proof. The statement is clearly true for $m = 1$ and 5. Every positive integer $m \geq 6$ has one of six forms: $6k$, $6k+1$, $6k+2$, $6k+3$, $6k+4$, or $6k+5$ for some positive integer k. If m is not divisible by 2 or 3, then $m = 6k+1$ or $m = 6k+5$ since each of $6k$, $6k+2$, $6k+3$, and $6k+4$ is divisible by 2 or by 3. For the case $m = 6k+1$ setting $n = 3k$ in Lemma 2.4 yields $m^2 = [2(3k)+1]^2 = 8T_{3k}+1$. But from Theorem 2.2 we have $T_{3k} = 6T_k + 3T_{k-1}$ and hence $(6k+1)^2 = 24(2T_k + T_{k-1}) + 1$. For the case $m = 6k+5$ letting $n = 3k+2$ in Lemma 2.4 yields $m^2 = [2(3k+2)+1]^2 = 8T_{3(k+1)-1} + 1$. But from Theorem 2.2 we have $T_{3(k+1)-1} = 6T_k + 3T_{k+1}$ and hence $(6k+5)^2 = 24(2T_k + T_{k+1}) + 1$. Thus when m is not divisible by 2 or 3, $m^2 \equiv 1 \pmod{24}$. □

In Example 1.8 we saw an example of *multi-polygonal numbers*, e.g., infinitely many numbers that are simultaneously square and triangular. Here is another proof of that result.

Theorem 2.6. *Infinitely many square triangular numbers exist.*

Proof. Setting $k = 8T_n$ in $T_k = \frac{k(k+1)}{2}$ yields $T_{8T_n} = \frac{8T_n(8T_n+1)}{2} = 4T_n(2n+1)^2$. So if T_n is a square, so is T_{8T_n}. Since $T_1 = 1^2$, we have the following infinite sequence of square triangular numbers: $T_1 = 1^2$, $T_8 = 6^2, T_{288} = 204^2, \ldots$. □

However, the above proof only shows that there are infinitely many *even* square triangular numbers. There are also infinitely many *odd* square triangular numbers, such as $T_{49} = 35^2$ and $T_{1681} = 1189^2$, but that requires a different proof. See Example 3.8 in the next chapter.

In Section 1.3 we illustrated how pentagonal and triangular numbers are related, e.g., $P_n = T_{2n-1} - T_{n-1}$. That relationship leads to the next result.

Theorem 2.7. *For* $n \geq 1$ *we have* $P_n \equiv n \pmod 3$.

Proof. The result is clearly true for $P_1 = 1$. In Figure 2.5 we see that $P_n = 3T_{n-1} + n$ for $n \geq 2$ (illustrated for $n = 8$). □

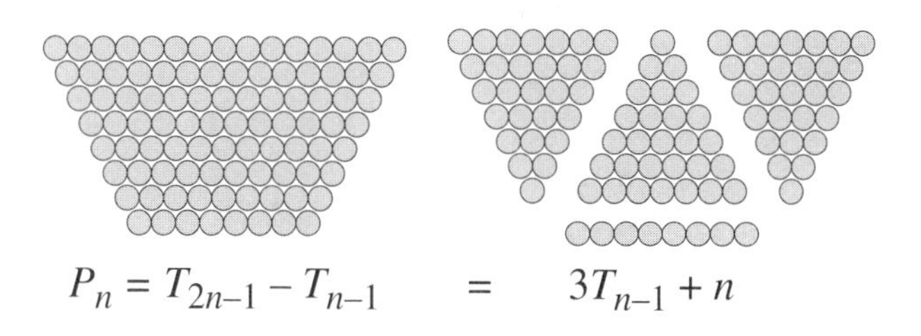

FIGURE 2.5

The fact that every polygonal number is a sum of triangular numbers enables us to use the congruence results for triangular numbers in Section 2.2 to evaluate similar congruences for n-gonal numbers with $n \geq 4$, as in the following example.

Example 2.1. *Heptagonal numbers modulo* 3. From Section 1.1 the kth heptagonal number is $\mathrm{Hep}_k = T_k + 4T_{k-1}$. Thus Theorem 2.2 yields

$$\begin{aligned}
\mathrm{Hep}_{3k-1} &= T_{3k-1} + 4T_{3k-2} = 3T_k + 42T_{k-1} + 4,\\
\mathrm{Hep}_{3k} &= T_{3k} + 4T_{3k-1} = 18T_k + 27T_{k-1},\\
\mathrm{Hep}_{3k+1} &= T_{3k+1} + 4T_{3k} = 33T_k + 12T_{k-1} + 1,
\end{aligned}$$

which proves

$$\mathrm{Hep}_n \equiv \begin{cases} 0 \pmod 3, & n \equiv 0 \pmod 3,\\ 1 \pmod 3, & n \not\equiv 0 \pmod 3. \end{cases}$$ □

Integer cubes are also related to triangular numbers, as we show in the following theorem.

Theorem 2.8. *For* $n \geq 1$ *we have* $(n+1)^3 - n^3 = 6T_n + 1$, *so that* $(n+1)^3 - n^3 \equiv 1 \pmod 6$.

Proof. In the first row of images in Figure 2.6 we see the difference of consecutive cubes as a "shell" of unit cubes. In the second row of images we replace each unit cube by a small sphere, and project that array of spheres onto the plane as a hexagonal array at the right end of the second row consisting of a central black sphere surrounded by six triangular arrays of gray spheres. □

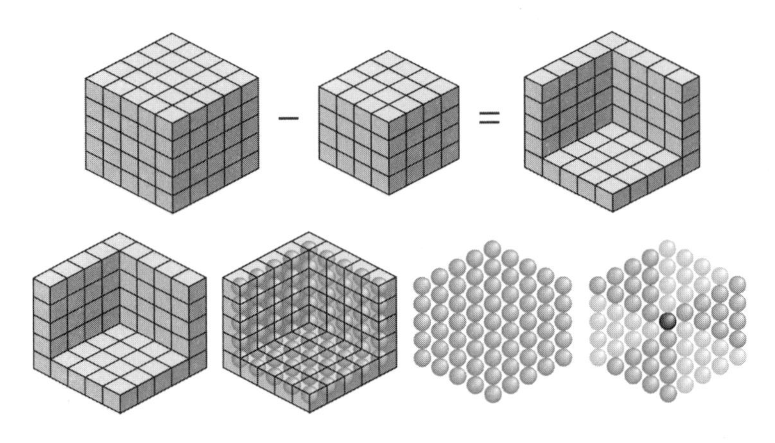

FIGURE 2.6

In Exercise 3.13 in the next chapter you can study the existence of consecutive cubes whose difference is a square or, equivalently, if $6T_n+1$ can equal a square.

2.3. Fermat's little theorem

The statement in the following theorem first appeared in a letter Pierre de Fermat wrote to Bernhard Frénicle de Bessy (1605–1675) in 1640. It is called his "little" theorem to distinguish it from his "last" or "great" theorem, the one about the lack of solutions to $x^n + y^n = z^n$ pictured on the postage stamp with Fermat at the end of Section 1.1.

There are a variety of proofs of this theorem. Ours is a combinatorial proof due to Solomon Golomb (1932–2016) [Golomb, 1965; Andrews, 1971].

Fermat's Little Theorem 2.9. *If p is prime and n is a positive integer, then $n^p \equiv n \pmod{p}$.*

Proof. We shall show that $p|(n^p - n)$. Suppose we have a collection of beads in n different colors, and we wish to make necklaces (or bracelets) by stringing together exactly p beads. Our first step is to put p beads on a string. Since each bead can be chosen in n ways, there are n^p possible strings of p beads. Exactly n of these strings will be monochromatic, and we discard these strings, leaving $n^p - n$ strings, each of which has beads of two or more colors.

Next we join seamlessly the ends of each string to form a necklace. However, many of the resulting necklaces will be indistinguishable, as illustrated in Figure 2.7 when $p = 5$ and $n = 3$.

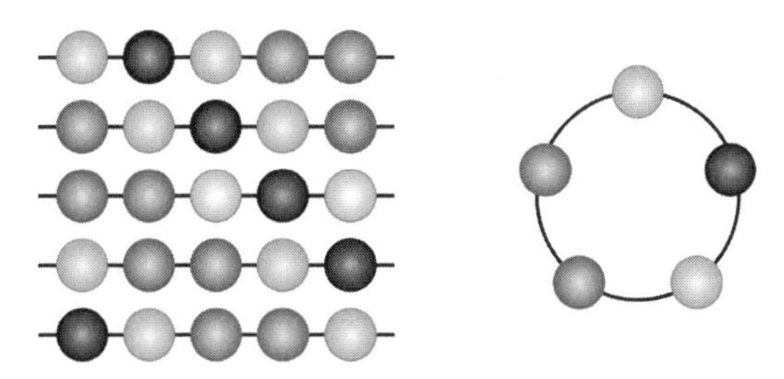

FIGURE 2.7

So for each string S in the collection, there will be another $p - 1$ strings that yield a necklace indistinguishable from S. Thus the number of *distinguishable* necklaces is $(n^p - n)/p$, which must be an integer. □

Where in the proof did we use the hypothesis that p is prime? It was in the last step where we asserted that for each string there were $p - 1$ additional strings that yield an indistinguishable necklace. If p is not prime, this does not hold. For example, let p be the composite number 6, and consider the strings in Figure 2.8.

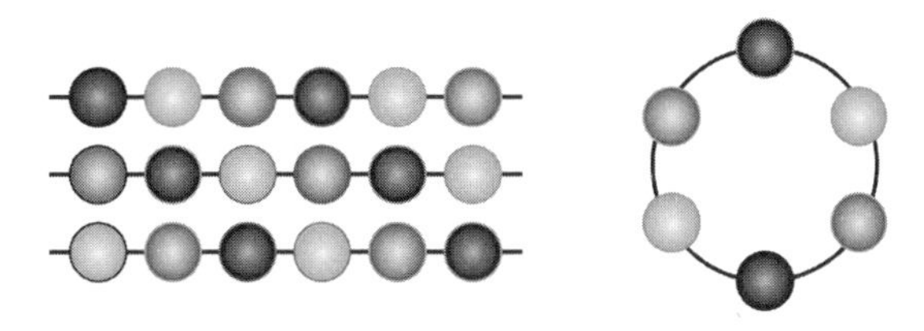

FIGURE 2.8

Here a set of three beads is repeated twice, and so that there are only $6/2 - 1 = 2$ additional strings yielding the same necklace.

Fermat's little theorem is often stated in the form of the following corollary.

Corollary 2.10. *If p is prime and n is a positive integer not divisible by p, then $n^{p-1} \equiv 1 \pmod{p}$.*

Proof. From the little theorem we have $p|(n^p - n)$ or, equivalently, $p|n(n^{p-1} - 1)$. Since the prime p does not divide n, it divides $n^{p-1} - 1$ and the conclusion follows. □

2.4. Wilson's theorem

In 1770 the English mathematician Edward Waring (1734–1798) published without proof a conjecture (the statement in the next theorem) of one of his students John Wilson (1741–1793). In the same year Lagrange gave the first proof. However, Gottfried Leibniz knew of the result before either Waring or Wilson was born. There are many proofs of the theorem; the following one is from [Carmichael, 1914; Andrews, 1971].

Wilson's Theorem 2.11. *If p is prime, then* $(p-1)! \equiv -1 \pmod{p}$.

Proof. We shall show that $p|[(p-1)!+1]$. The theorem is clearly true for $p = 2$ and $p = 3$ ($2|(1!+1)$ and $3|(2!+1)$), so assume that $p \geq 5$. Now choose p points on a circle, one directly above the center and the remaining $p-1$ equally spaced around the circle. These points define the vertices of a *regular p-gon* (a polygon with p congruent sides and angles). We now construct *stellated p-gons* by connecting the vertices in every possible order (we allow edges to cross one another). The number of *distinct* stellated p-gons is $\frac{1}{2}(p-1)\cdot(p-2)\cdots 2\cdot 1 = (p-1)!/2$. Since the first vertex is the vertex directly above the center, the second can be chosen in $p-1$ ways, the third in $p-2$ ways, and so on to the last in one way; we divide by 2 since we can choose one or the other of the two edges that meet at the first vertex as the first edge (i.e., there are two ways in which to join the vertices to form the stellated p-gon). See Figure 2.9 for the $4!/2 = 12$ stellated pentagons.

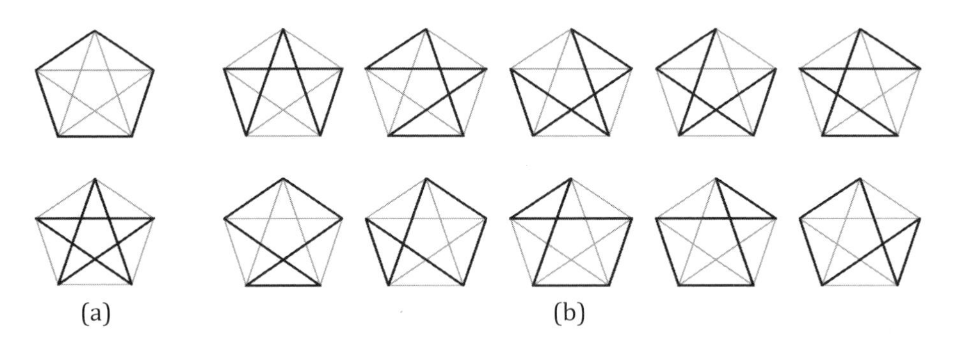

FIGURE 2.9

Of the total number of stellated p-gons, $(p-1)/2$ are *regular* stellated p-gons, in that the angles and side lengths are equal. These are constructed by connecting every kth vertex of the p-gon for $1 \leq k \leq (p-1)/2$. For $p = 5$ these are the two regular stellated pentagons in

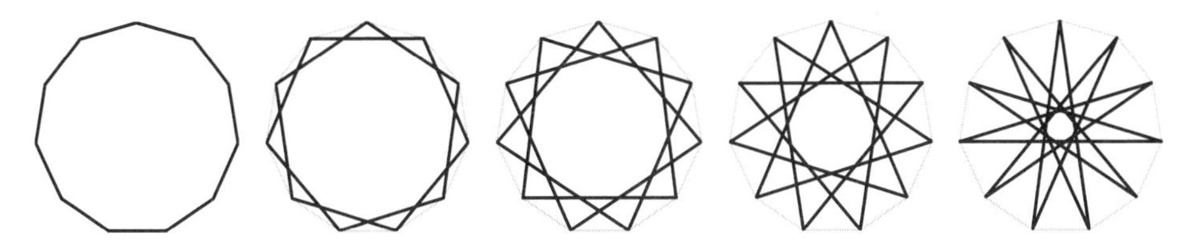

FIGURE 2.10

Figure 2.9(a), and for $p = 11$ the five regular stellated 11-gons are illustrated in Figure 2.10.

The $[(p-1)!/2]-[(p-1)/2]$ non-regular stellated p-gons fall into subsets of size p, since for each such p-gon P there are $p-1$ others which can be obtained from P by rotating P through an angle of $(360/p)^\circ$. When $p = 5$ there are two such subsets, the two rows of stellated pentagons in Figure 2.9(b). Thus $[(p-1)!-(p-1)]/2p$ is an integer, so that p divides $(p-1)!+1$. □

2.5. Exercises

2.1 Prove that in base 9, every integer consisting solely of 1's is a triangular number, e.g., $11_9 = 9+1 = 10 = T_4$, $111_9 = 81+9+1 = 91 = T_{13}$, etc. (here the subscript on an integer indicates a base other than 10). [Hint: See Exercise 1.2.]

2.2 Examination of the columns of Table 1.1 leads to the observation that when k is odd, the kth n-gonal number is congruent to 0 modulo k. Prove that this is indeed the case.

2.3 Give a third proof of Theorem 2.2 by proving the following three identities:
(a) $T_{3k-1} = 3(T_{2k} - T_{k+1}) + T_2$; (b) $T_{3k} = 3(T_{2k} - T_k)$; and (c) $T_{3k+1} = 3(T_{2k+1} - T_{k+1}) + T_1$.

2.4 Prove that there exist infinitely many pairs (a, b) of integers such that for every positive integer t the number $at + b$ is a triangular number if and only if t is a triangular number. [This is problem B6 from the 1988 William Lowell Putnam Competition.]

2.5 Prove the following corollary to Fermat's little theorem: *If p is an odd prime and n is a positive integer, then* $n^p \equiv n \pmod{2p}$.

2.6 In Theorem 2.8 we proved that the difference of consecutive cubes is congruent to 1 modulo 6. Now prove that each cube n^3 is congruent to n modulo 6.

2.7 Let n be a positive integer. Prove that in base 10, n and n^5 have the same units digit, e.g., $\underline{7}^5 = 1680\underline{7}$ and $1\underline{2}^5 = 24883\underline{2}$. [Hint: Two numbers have the same units digit if they are congruent modulo 10.]

2.8 In Example 1.10 we mention Fermat primes. More generally, a *Fermat number* is an integer of the form $f_n = 2^{2^n} + 1$ for $n \geq 0$. The first few Fermat numbers are $f_0 = 3$, $f_1 = 5$, $f_2 = 17$, $f_3 = 257$, $f_4 = 65,537$, and $f_5 = 4,294,967,297$. Prove that for $n \geq 2$ (a) the units digit of f_n is 7, and (b) the tens digit of f_n is odd.

2.9 Show that the nth Fermat number satisfies $f_n \equiv 5 \pmod{12}$ for $n \geq 1$.

2.10 Observe that $1 + 3 + 6 + 10 + 15 + 21 + 28 = 84 = 3 \cdot 28$. Are there other instances where $\sum_{k=1}^{n} T_k \equiv 0 \pmod{T_n}$?

2.11 For what values of n is $P_n \equiv 0 \pmod{5}$?

2.12 Consider the determinant $|a_{ij}|$ of order 100 with $a_{ij} = i \times j$. Prove that if the absolute value of each of the 100! terms in the expansion of this determinant is divided by 101, then the remainder in each case is 1. [This is problem B1 from the 1957 William Lowell Putnam Competition.]

2.13 Show that p is the smallest prime that divides $(p-1)!+1$.

CHAPTER 3

Diophantine Equations

Equations are important to me, because politics is for the present, but an equation is something for eternity.

Albert Einstein

I do believe in simplicity. When the mathematician would solve a difficult problem he first frees the equation from all encumbrances, and reduces it to its simplest terms.

Henry David Thoreau

A *Diophantine equation* is an equation with integer coefficients in which the unknowns are also integers. They are named after Diophantus of Alexandria (circa 200–284 CE), who made an extensive study of equations whose coefficients and solutions were integers or rational numbers. In this chapter we discuss and illustrate linear Diophantine equations such as $ax + by = c$ (equivalent to the congruence $ax \equiv c \pmod{b}$) in Sections 3.2 and 3.3; quadratic Diophantine equations of the form $x^2 - dy^2 = 1$ in Sections 3.4 through 3.6; and quadratic Diophantine equations of the form $a^2 + b^2 = c^2$ in the next chapter.

Example 3.1. On the left in Figure 3.1 we see a graph of the linear equation $2x - 3y = 1$ with some of its infinitely many integer solutions marked with •; a graph of the quadratic equation $x^2 - 2y^2 = 1$ with some of its infinitely many integer solutions in the center; and a graph of the quadratic equation $x^2 + y^2 = 13$ with all eight of its integer solutions on the right. □

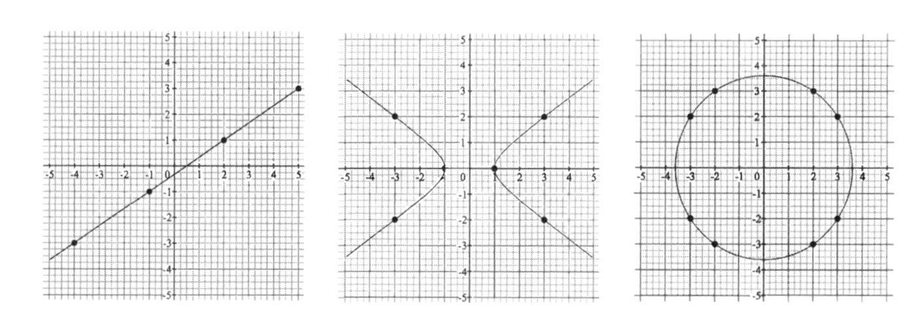

FIGURE 3.1

Diophantine equations and David Hilbert

On August 8, 1900, David Hilbert (1862–1943) presented an address at the Second International Congress of Mathematicians in Paris. In this address, Hilbert presented 23 problems (in the printed version, but only 10 in the oral presentation), the solutions to which he believed would profoundly impact the course of mathematical research in the twentieth century. In his 10th problem, Hilbert asked: *Does there exist an algorithm to determine whether a given polynomial Diophantine equation has a solution in integers*? This question was answered in the negative (no such algorithm exists) in the work of Martin Davis, Hilary Putnam, Julia Robinson, and Yuri Matiyasevich over a span of more than 20 years, ending in 1970.

David Hilbert

The above photograph was taken in 1912 for postcards of faculty members at the University of Göttingen.

We begin in the next section with a way to illustrate some properties of squares using triangles.

3.1. Triangles and squares

In some instances it will be useful to use equilateral triangles to represent integral squares. To do so, we first recall that every square is the sum of consecutive odd numbers, e.g.,

$$n^2 = 1 + 3 + 5 + \cdots + (2n - 1),$$

as illustrated in Figure 1.1(b) and in Figure 3.2(a) for $n = 8$ (count the dots in each of the L-shaped regions).

This enables us to represent n^2 using an equilateral triangle subdivided into n rows containing 1, 3, 5, ..., $2n - 1$ identical smaller equilateral triangles (Δ and ∇), as seen in Figure 3.2(b). As an application, we color the small triangles as shown in Figure 3.2(c) to illustrate the identity $T_n + T_{n-1} = n^2$ (for $n = 8$) with T_n small gray triangles and T_{n-1} small white triangles.

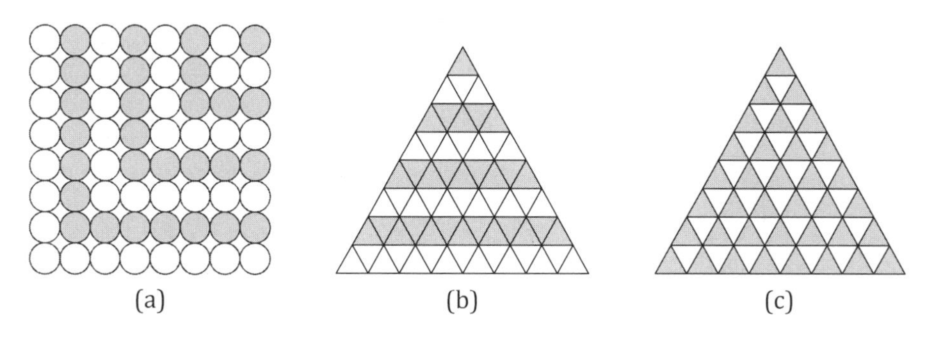

FIGURE 3.2

Example 3.2. *Squares modulo* 3. The $n = 4$ row in Table 1.1 yields the hypothesis that every square is either a multiple of 3 or 1 more than a multiple of 3. In Theorem 2.2 we used a triangular array of dots to evaluate triangular numbers modulo 3. With triangular diagrams similar to those in Figure 3.2 we can evaluate squares modulo 3:

$$n^2 \equiv \begin{cases} 0 \pmod 3, & n \equiv 0 \pmod 3, \\ 1 \pmod 3, & n \not\equiv 0 \pmod 3. \end{cases}$$

See Figure 3.3, and note that a set of small triangles in the shape of a trapezoid represents the difference of two squares. The figure is drawn for $k = 3$, representing n^2 for n equal to 8, 9, and 10. □

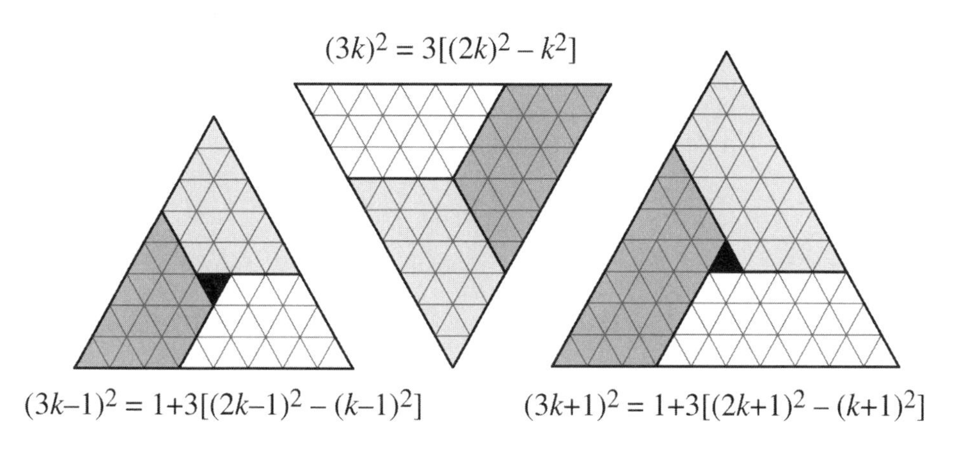

FIGURE 3.3

Example 3.3. *Fermat's little theorem for* $p = 3$. Figure 3.3 also illustrates the little theorem for $p = 3$ in the form expressed in Corollary 2.10, i.e., if n is an integer not divisible by 3, then $n^2 \equiv 1 \pmod 3$. □

Example 3.4. *An identity for seven squares.* For positive integers a, b, and c, we have

$$a^2 + b^2 + c^2 + (a + b + c)^2 = (a + b)^2 + (a + c)^2 + (b + c)^2.$$

In Figure 3.4 we use the inclusion-exclusion principle to count the small Δ and ∇ triangles, obtaining

$$(a + b + c)^2 = (a + b)^2 + (a + c)^2 + (b + c)^2 - a^2 - b^2 - c^2,$$

from which the desired result follows. Simple algebra shows that the identity actually holds for all real numbers.

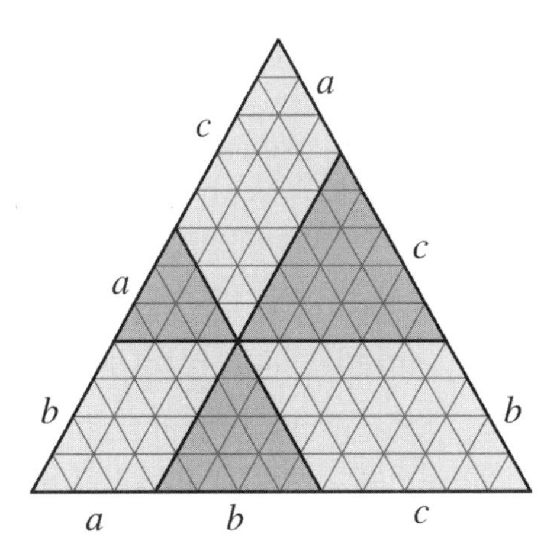

FIGURE 3.4

In the figure $(a, b, c) = (3, 4, 5)$, which yields $3^2 + 4^2 + 5^2 + 12^2 = 7^2 + 8^2 + 9^2$. The change of variables

$$(a, b, c) = (-x + y + z, x - y + z, x + y - z)$$

yields

$$(x+y+z)^2+(x+y-z)^2+(x-y+z)^2+(-x+y+z)^2 = 4(x^2+y^2+z^2).$$

Other identities for squares can be obtained in a similar fashion. □

3.2. Linear Diophantine equations

A *linear Diophantine equation* is an equation of the form $ax + by = c$, where a, b, and c are given integers and x and y are integer-valued unknowns. The task is to *solve* the equation, finding integer values of x and y so that $ax + by = c$.

A few observations:

1. Let $d = \gcd(a, b)$, the greatest common divisor of a and b. Hence d is always a factor of $ax + by$. So if d is not a factor of c, then $ax + by = c$ is not solvable in integers.
2. Assume d is a factor of c. Then we can divide through by d to obtain $\frac{a}{d}x + \frac{b}{d}y = \frac{c}{d}$, where $\frac{a}{d}$, $\frac{b}{d}$, and $\frac{c}{d}$ are integers and $\gcd(\frac{a}{d}, \frac{b}{d}) = 1$. Furthermore, any integer values of x and y that satisfy $\frac{a}{d}x + \frac{b}{d}y = \frac{c}{d}$ also satisfy $ax + by = c$, and vice versa.
3. Hence solving $ax + by = c$ is reduced to the case of solving the equation when $\gcd(a, b) = 1$. But as a consequence of the Euclidean algorithm for finding $\gcd(a, b)$, there exist integers r and s such that $ar + bs = 1$, hence $ax + by = c$ has a solution $x_0 = rc$ and $y_0 = sc$.

The Euclidean algorithm

The Euclidean algorithm is an arithmetic procedure for finding the greatest common divisor of two positive integers a and b, so named since it appears in Book VII of Euclid's *Elements*. Here is a geometric version [Křížek et al., 2001; Walser, 2001]. Draw an a-by-b $(a > b)$ rectangle, as shown below. Cut off as many squares with side b as possible, until a rectangle remains. Again cut off squares from the new rectangle, and continue until no rectangle remains. The side length of the smallest square is then $\gcd(a, b)$. The figure below illustrates the procedure with $(a, b) = (64, 27)$ and $\gcd(64, 27) = 1$.

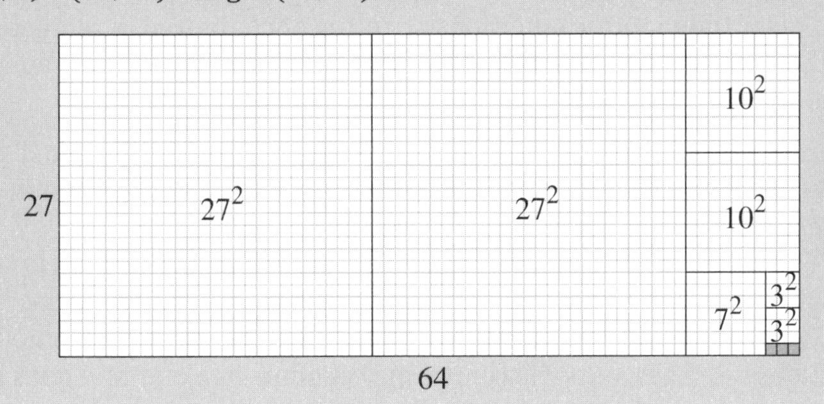

A method for finding all solutions of $ax + by = c$ (when $\gcd(a, b) = 1$) is based on the following lemma.

Lemma 3.1. *If a, b, x, and y are integers, then $ax + by = a(x - b) + b(y + a)$.*

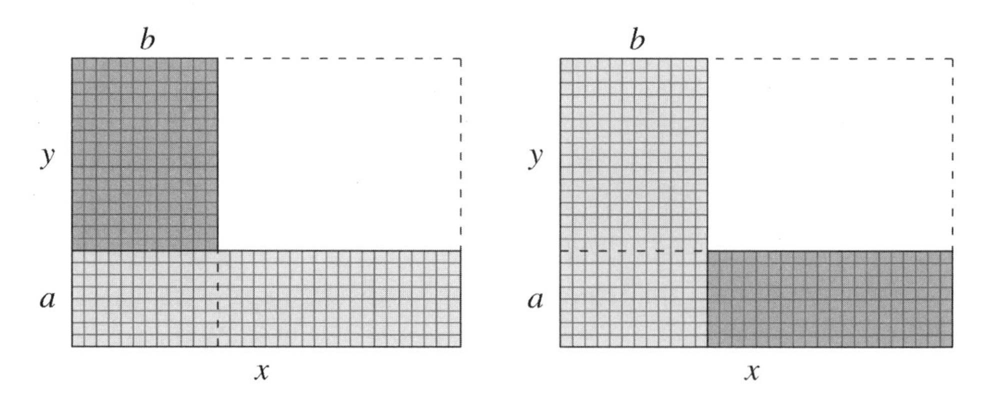

FIGURE 3.5. $ax + by = a(x - b) + b(y + a)$

Proof. The result follows from simple algebra. In Figure 3.5 we illustrate the result for positive integers by counting the number of small squares in two different ways as indicated by the shading. □

Similar algebra establishes $ax + by = a(x - kb) + b(y + ka)$ for any integer k. Alternatively, you can also derive this from the formula in the lemma by replacing x by $x \pm b$, $x \pm 2b$, etc., and similarly replacing y by $y \pm a$, $y \pm 2a$, etc. This leads to the following theorem.

Theorem 3.2. *Suppose a and b are non-zero integers with* $\gcd(a, b) = 1$, *and that x_0, y_0 is a solution to $ax+by = c$. Then all solutions to $ax+by = c$ are given by $x = x_0 - kb$, $y = y_0 + ka$ for some integer k.*

The proof that all solutions have the stated form can be found in most number theory textbooks, e.g., [Marshall et al., 2007].

Linear Diophantine equations are often encountered in recreational mathematics puzzles, such as the one in the following example adapted from [Gardner, 2001].

Example 3.5. Three sailors and a monkey were shipwrecked on a desert island and spent a day gathering coconuts, as illustrated in a Javanese drawing in Figure 3.6.

They piled the coconuts together, and agreed to divide them up after a nights rest. During the night one sailor wakes up, divides the coconuts into three equal piles with one left over, which he gives to the monkey. He hides his share, puts the remaining coconuts in a pile, and goes back to sleep. The second sailor does the same thing, as does the third. In the morning the sailors divide the remaining pile of coconuts, with one left over for the monkey. How many coconuts were there in the original pile?

Let x denote the number of coconuts in the original pile, a, b, and c the number taken by each sailor during the night, and y the number

FIGURE 3.6

each received in the morning. Then $x = 3a+1, 2a = 3b+1, 2b = 3c+1$, and $2c = 3y+1$. Eliminating a, b, and c yields the Diophantine equation $8x - 81y = 65$. Since $(x, y) = (-10, -1)$ is a solution to $8x - 81y = 1$, $(x_0, y_0) = (-650, -65)$ is a solution to $8x - 81y = 65$. Hence the general solution is $(x, y) = (-650 + 81k, -65 + 8k)$. For $k \geq 9$ we have the positive solutions $(x, y) = (79, 7), (160, 15), (241, 23)$, etc. □

This problem originally appeared in the October 9, 1926 issue of *The Saturday Evening Post*, with five sailors rather than three [Gardner, 2001]. You are invited to solve the original version in Exercise 3.1.

3.3. Linear congruences and the Chinese remainder theorem

Knowing how to solve a linear Diophantine equation enables one to solve a *linear congruence*, a congruence of the form $ax \equiv c \pmod{b}$ for x, since the congruence is equivalent to the linear Diophantine equation $ax + by = c$. So we now consider solving a set of simultaneous linear congruences, e.g., finding values of x that satisfy $x \equiv c_i \pmod{m_i}$ for $1 \leq i \leq k$. The key to solving such systems is the following theorem, whose proof can be found in a number theory textbook.

The Chinese Remainder Theorem 3.3. *The system of k linear congruences $x \equiv c_i \pmod{m_i}$ for $1 \leq i \leq k$, where the moduli are relatively prime in pairs, has a unique solution modulo $m_1 m_2 \cdots m_k$.*

Note that the theorem tells us when the system has solutions, but does not provide a procedure for finding them. The name of the theorem honors the contributions of early Chinese mathematicians to the theory of congruences. There are a variety of methods for finding solutions to systems of linear congruences. The following two examples illustrate a tabular procedure applicable to systems with relatively small moduli.

Example 3.6. Consider a system of two congruences: $x \equiv 2 \pmod 3$ and $x \equiv 3 \pmod 5$. The Chinese remainder theorem guarantees the existence of a solution mod 15. To find it, we set up a 3-by-5 table with row labels mod 3 and column labels mod 5. Highlighting the row and column corresponding to the given congruences identifies the cell that will contain the solution; here it is cell (2,3). See Table 3.1.

TABLE 3.1

		mod 5				
		0	1	2	3	4
	0	0	6		3	
mod 3	1		1	7		4
	2	5		2	8	

We now successively enter the numbers 0, 1, 2, ..., starting in the (0,0) cell and proceeding down the diagonal until we reach an edge of the table. When we reach the bottom edge we jump up to the top edge, and when we reach the right edge we revert to the left edge of the table, as if the table were the map of a torus, where the top and bottom edges of the table are adjacent, as are the right and left edges. See Figure 3.7.

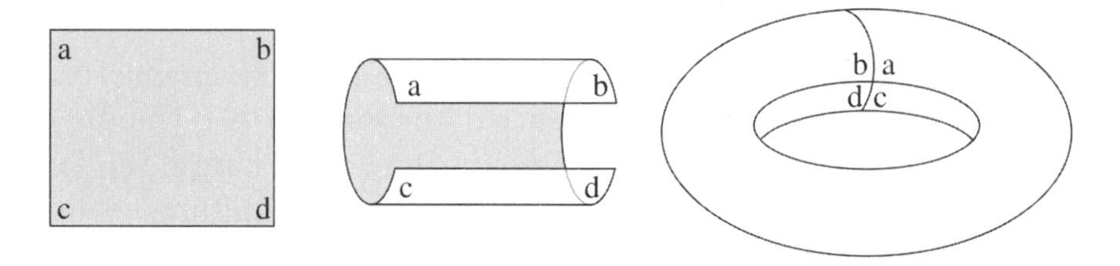

FIGURE 3.7

When we reach cell (2,3) in the table, we stop entering numbers as we have found the solution: $x \equiv 8 \pmod{15}$. □

Example 3.7. *Sun Tzu's puzzle.* Perhaps the earliest use of the Chinese remainder theorem appears in Problem 26 of Chapter 3 in *Sun Tzu Suan Ching* (*Sun Tzu's Mathematical Manual*), circa 250 CE (see Figure 3.8):

> "Suppose we have an unknown number of objects. When counted in threes, 2 are left over, when counted in fives, 3 are left over, and when counted in sevens, 2 are left over. How many objects are there?"

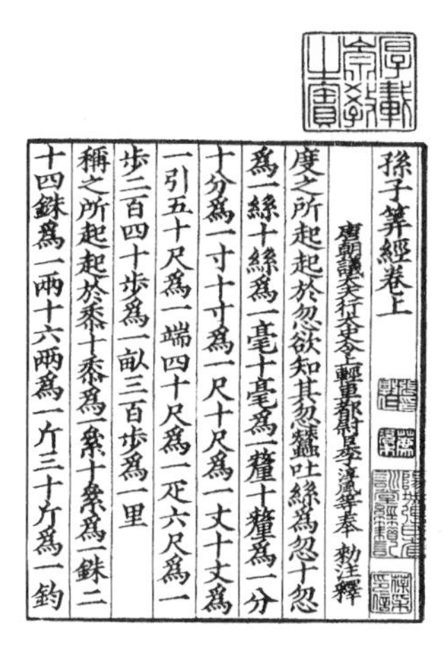

孫子算經卷上
唐朝議大夫行太史令上輕車都尉臣李淳風等奉勅注釋
度之所起起於忽欲知其忽蠶吐絲爲忽十忽
爲一絲十絲爲一毫十毫爲一釐十釐爲一分
十分爲一寸十寸爲一尺十尺爲一丈十丈爲
一引五十尺爲一端四十尺爲一疋六尺爲一
步二百四十步爲一畝三百步爲一里
稱之所起起於黍十黍爲一絫十絫爲一銖二
十四銖爲一兩十六兩爲一斤三十斤爲一鈞

FIGURE 3.8. Facsimile of Qing dynasty edition of *Sun Tzu's Mathematical Manual*

Solving Sun Tzu's puzzle requires solving the system

$$x \equiv 2 \pmod{3},$$
$$x \equiv 3 \pmod{5},$$
$$x \equiv 2 \pmod{7}.$$

We can replace the first two equations with $x \equiv 8 \pmod{15}$ from Example 3.6, and solve the system

$$x \equiv 8 \pmod{15},$$
$$x \equiv 2 \pmod{7},$$

using our tabular method, as illustrated in Table 3.2.

When we reach cell (2,8) we stop entering numbers, as we have found the solution $x \equiv 23 \pmod{105}$. So the number of objects in Sun Tzu's puzzle is one of the numbers 23, 128, 233, 338, ... □

TABLE 3.2

		mod 15														
		0	1	2	3	4	5	6	7	8	9	10	11	12	13	14
mod 7	0	0						21	7							14
	1	15	1						22	8						
	2		16	2						23	9					
	3			17	3							10				
	4				18	4							11			
	5					19	5							12		
	6						20	6							13	

3.4. The Pell equation $x^2 - 2y^2 = 1$

Some of the most often studied quadratic Diophantine equations are the so-called *Pell equations* $x^2 - dy^2 = c$, where c is an integer and d is a positive non-square integer. Such equations may have infinitely many solutions. When d is a negative integer or a positive integral square, the Pell equation has at most finitely many solutions.

These equations were mistakenly attributed to a 17th century Englishman John Pell by Leonard Euler in 1730. This was unfortunate, for as the mathematical historian E. T. Bell [Bell, 1990] writes, "Pell mathematically was a nonentity, and humanly an egregious fraud. It is long past time that his name be dropped from the textbooks."

Some authors refer to these equations as *Fermat equations*, since Pierre de Fermat initiated their modern history with a letter in 1657 to several mathematicians challenging them to solve $x^2 - 61y^2 = 1$ in integers. However, these equations were studied in India by Brahmagupta (circa 598–665 CE) and others long before Fermat.

The standard way of solving Pell equations is via the theory of continued fractions. Our approach to solving $x^2 - 2y^2 = 1$ is more combinatorial and is based on counting the small squares in Figure 3.9.

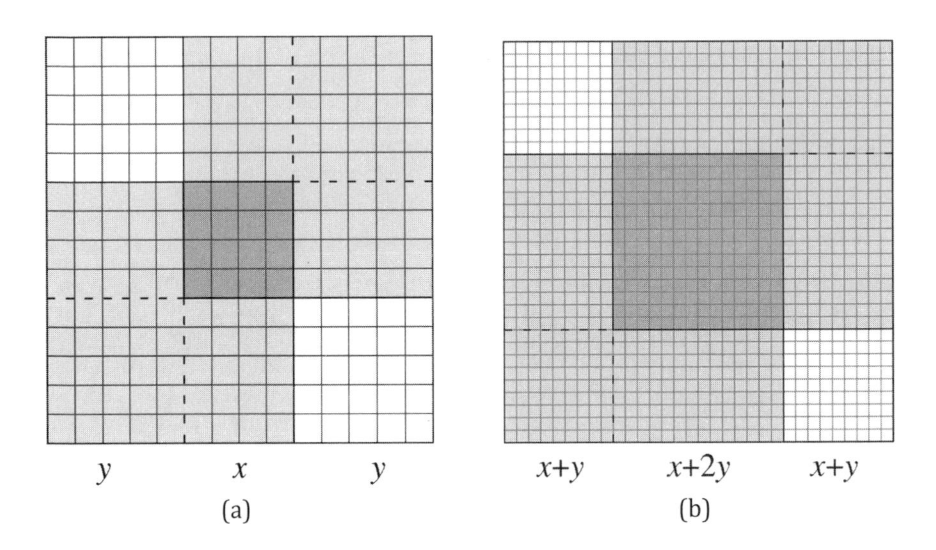

FIGURE 3.9

Counting the number of small squares in Figure 3.9(a) using the inclusion-exclusion principle yields $(x + 2y)^2 = 2(x + y)^2 - x^2 + 2y^2$, from which it follows that $x^2 - 2y^2 = 2(x + y)^2 - (x + 2y)^2$. Similarly counting the small squares in Figure 3.9(b) yields $(3x+4y)^2 = 2(2x+3y)^2 - (x+2y)^2 + 2(x+y)^2$, so that $2(x+y)^2 - (x+2y)^2 = (3x+4y)^2 - 2(2x+3y)^2$. Hence $x^2 - 2y^2 = (3x+4y)^2 - 2(2x+3y)^2$ and thus

$$x^2 - 2y^2 = 1 \text{ if and only if } (3x+4y)^2 - 2\,(2x+3y)^2 = 1. \tag{3.1}$$

We now use (3.1) to generate a sequence (x_n, y_n) satisfying $x_n^2 - 2y_n^2 = 1$ using the recursions $x_{n+1} = 3x_n + 4y_n$, $y_{n+1} = 2x_n + 3y_n$ and starting with an initial solution $(x_0, y_0) = (1, 0)$. The results are shown in Table 3.3 for n from 0 to 5.

TABLE 3.3

n	0	1	2	3	4	5
x_n	1	3	17	99	577	3363
y_n	0	2	12	70	408	2378

We now find explicit expressions for the solutions x_n and y_n. The recursions $x_{n+1} = 3x_n + 4y_n$ and $y_{n+1} = 2x_n + 3y_n$ imply that

$$\begin{aligned} x_{n+1} &= 3x_n + 4(2x_{n-1} + 3y_{n-1}) \\ &= 3x_n + 8x_{n-1} + 3(x_n - 3x_{n-1}) \\ &= 6x_n - x_{n-1}, \\ y_{n+1} &= 2(3x_{n-1} + 4y_{n-1}) + 3y_n \\ &= 3\,(y_n - 3y_{n-1}) + 8y_{n-1} + 3y_n \\ &= 6y_n - y_{n-1}, \end{aligned}$$

i.e., $x_{n+1} = 6x_n - x_{n-1}$ and $y_{n+1} = 6y_n - y_{n-1}$. Evaluating the generating functions $G_x(t) = \sum_{n=0}^{\infty} x_n t^n$ and $G_y(t) = \sum_{n=0}^{\infty} y_n t^n$ for the sequences $\{x_n\}$ and $\{y_n\}$ yields

$$G_x(t) = (1-3t)/(1-6t+t^2) \quad \text{and} \quad G_y(t) = 2t/(1-6t+t^2),$$

from which we obtain

$$x_n = \frac{1}{2}[(3+2\sqrt{2})^n + (3-2\sqrt{2})^n]$$

and

$$y_n = \frac{1}{2\sqrt{2}}[(3+2\sqrt{2})^n - (3-2\sqrt{2})^n]$$

via partial fraction decompositions and geometric series.

Example 3.8. *Square triangular numbers.* In Example 1.8 we saw that infinitely many square triangular numbers exist. We obtained the same result in Theorem 2.6, but the proof of that theorem generated only *even* square triangular numbers. In this example we generate the infinite sequence of square triangular numbers $T_m = k^2$ using the Pell equation $x^2 - 2y^2 = 1$. Completing the square for both m and k shows that $T_m = k^2$ is equivalent to $(2m+1)^2 - 2(2k)^2 = 1$, hence each solution (x_n, y_n) in Table 3.3 yields a square triangular number. For example $(x_2, y_2) = (17, 12)$ yields $(m, k) = (8, 6)$ so that $T_8 = 36 = 6^2$, $(x_3, y_3) = (99, 70)$ yields $(m, k) = (49, 35)$ so that $T_{49} = 1225 = 35^2$, $(x_4, y_4) = (577, 408)$ yields $(m, k) = (288, 204)$ so that $T_{288} = 41616 = 204^2$, and so on. □

Solutions to the Pell equation $x^2 - 2y^2 = c$ for another integer c are found similarly. If we have $x^2 - 2y^2 = -1$ we have the same recursion as above but with an initial solution $(x_0, y_0) = (1, 1)$, yielding the results in Table 3.4 for n from 0 to 5.

TABLE 3.4

n	0	1	2	3	4	5
x_n	1	7	41	239	1393	8119
y_n	1	5	29	169	985	5741

Example 3.9. *Oblong triangular numbers.* As we observed in Example 1.8, infinitely many positive integers are both oblong and triangular: $T_3 = 6 = 2 \cdot 3$ and $T_{20} = 210 = 14 \cdot 15$, for example. As in the preceding example, we can generate the sequence of oblong triangular numbers by considering $T_m = k(k+1)$ and completing the square to obtain $(2m+1)^2 - 2(2k+1)^2 = -1$. Thus the solutions (x_n, y_n) in Table 3.4 to the Pell equation $x^2 - 2y^2 = -1$ yield oblong triangular numbers. For example $(x_3, y_3) = (239, 169)$ yields $(m, k) = (119, 84)$, and $T_{119} = 7140 = 84 \cdot 85$. □

3.5. The Pell equation $x^2 - 3y^2 = 1$

To solve this Pell equation, we first count the small triangles in Figure 3.10 using the inclusion-exclusion principle, recalling that each large triangle contains a square number of small triangles, as illustrated in Figure 3.2(b).

This yields $(2x+3y)^2 = 3(x+2y)^2 + x^2 - 3y^2$, and hence

$$x^2 - 3y^2 = 1 \text{ if and only if } (2x+3y)^2 - 3(x+2y)^2 = 1. \tag{3.2}$$

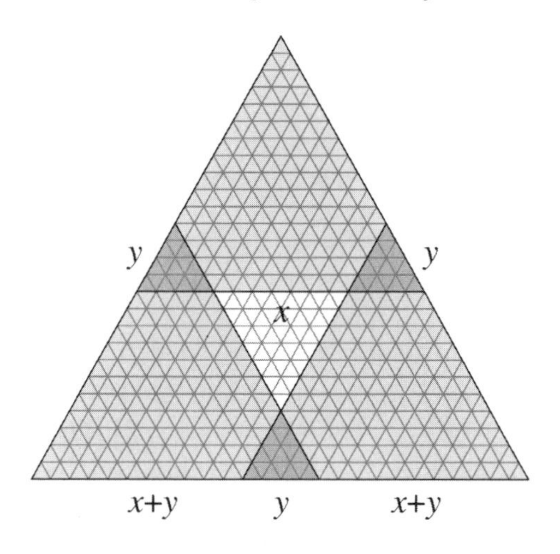

FIGURE 3.10

We now use (3.2) to generate a sequence (x_n, y_n) satisfying $x_n^2 - 3y_n^2 = 1$ using the recursions $x_{n+1} = 2x_n + 3y_n$, $y_{n+1} = x_n + 2y_n$ and starting with an initial solution $(x_0, y_0) = (1, 0)$, yielding the results in Table 3.5 for n from 0 to 6.

TABLE 3.5

n	0	1	2	3	4	5	6
x_n	1	2	7	26	97	362	1351
y_n	0	1	4	15	56	209	780

We now find explicit expressions for the solutions x_n and y_n. The recursions $x_{n+1} = 2x_n + 3y_n$ and $y_{n+1} = x_n + 2y_n$ imply that $x_{n+1} = 4x_n - x_{n-1}$ and $y_{n+1} = 4y_n - y_{n-1}$. Evaluating the generating functions $G_x(t) = \sum_{n=0}^{\infty} x_n t^n$ and $G_y(t) = \sum_{n=0}^{\infty} y_n t^n$ for the sequences $\{x_n\}$ and $\{y_n\}$ yields $G_x(t) = (1-2t)/(1-4t+t^2)$ and $G_y(t) = t/(1-4t+t^2)$, from which we obtain

$$x_n = \frac{1}{2}[(2+\sqrt{3})^n + (2-\sqrt{3})^n]$$

and

$$y_n = \frac{1}{2\sqrt{3}}[(2+\sqrt{3})^n - (2-\sqrt{3})^n]$$

via partial fraction decompositions and geometric series.

Example 3.10. *Almost equilateral Heronian triangles.* A *Heronian triangle* is one whose sides (a, b, c) and area K are integers (some sources replace "integers" with "rational numbers"). An *almost equilateral*

Heronian triangle is a Heronian triangle whose sides are consecutive integers, such as $(3,4,5)$ with area 6 and $(13,14,15)$ with area 84. Are there others? In answering, we recall that the area K of an arbitrary triangle (a,b,c) is given by *Heron's formula:*

$$K = \sqrt{s(s-a)(s-b)(s-c)},$$

where s denotes the *semiperimeter*, $s = (a+b+c)/2$.

Letting $(b-1, b, b+1)$ denote the sides of the triangle yields $s = 3b/2$ and $K^2 = (3b/2)(b/2)[(b-2)/2][(b+2)/2]$ or $16K^2 = 3b^2(b^2-4)$. Thus b must be even, so we set $b = 2x$, hence $K^2 = 3x^2(x^2-1)$. Then x^2-1 must be 3 times a square, so that $x^2-1 = 3y^2$, the Pell equation solved in this section. So pairs (x_n, y_n) in Table 3.5 yield triangles with sides $(2x_n-1, 2x_n, 2x_n+1)$ and area $K = 3x_ny_n$ (thus the altitude h to the even side $2x_n$ is the integer $h = 3y_n$). Hence there are infinitely many almost equilateral Heronian triangles. For example, $n = 3$ yields the triangle $(51, 52, 53)$ with area $K = 1170$. The *inradius* (the radius r of the inscribed circle) of an almost equilateral Heronian triangle is also an integer; in Exercise 3.16 you can prove that $r = y_n$. In Figure 3.11 we see an illustration of the almost equilateral Heronian triangle $(2x_n-1, 2x_n, 2x_n+1)$ and its altitude and inscribed circle for the case $(x_2, y_2) = (7,4)$. □

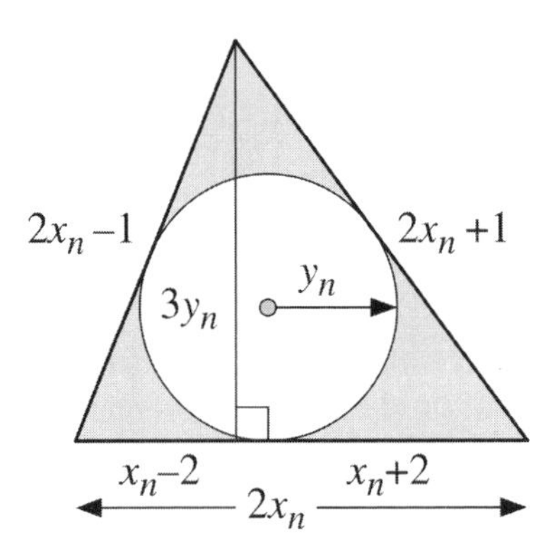

FIGURE 3.11

Solutions to the Pell equation $x^2 - 3y^2 = c$ for another integer c are found similarly. If we have $x^2 - 3y^2 = -2$ we have the same recursion as above but with an initial solution $(x_0, y_0) = (1,1)$, yielding the results in Table 3.6 for n from 0 to 6.

TABLE 3.6

n	0	1	2	3	4	5	6
x_n	1	5	19	71	265	989	3691
y_n	1	3	11	41	153	571	2131

Example 3.11. *Pentagonal triangular numbers.* In Chapter 1 we encountered numbers that were simultaneously pentagonal and triangular: $P_1 = 1 = T_1$ (which is rather uninteresting) and $P_{12} = 210 = T_{20}$, for example. Are there more? How many? As we did with square triangular numbers, we look to solve $P_m = T_k$, i.e., $m(3m-1)/2 = k(k+1)/2$ for m and k. Completing the square on m and k yields $(6m-1)^2 - 3(2k+1)^2 = -2$, the Pell equation $x^2 - 3y^2 = -2$ with $x = 6m - 1$ and $y = 2k + 1$. Hence the pairs (x_n, y_n) in Table 3.6 with $x_n \equiv -1 \pmod 6$ yield solutions, e.g., $(x_3, y_3) = (71, 41)$ yields $(m, k) = (12, 20)$ with $P_{12} = 210 = T_{20}$, $(x_5, y_5) = (989, 571)$ yields $(m, k) = (165, 285)$ with $P_{165} = 40755 = T_{285}$, etc. □

3.6. The Pell equations $x^2 - dy^2 = 1$

We now consider $x^2 - dy^2 = 1$, where d is positive and non-square. Note that $(x, y) = (1, 0)$ is a solution for all d, and the only solution when d is square. A proof that $x^2 - dy^2 = 1$ (for d positive and non-square) has infinitely many solutions in *positive* integers whenever it has at least one such solution is based on the following simple lemma.

Lemma 3.4. *For* $u, v > 0$, $(u+v)^2 = (u-v)^2 + 4uv$.

Proof. Without loss of generality assume $u \geq v$ and see Figure 3.12 for a visual proof. Of course the lemma can also be proved by elementary algebra. □

Theorem 3.5. *If* $x^2 - dy^2 = 1$, *where* d *is positive and non-square, has a solution in positive integers, then it has infinitely many.*

Proof. Set $u = x^2$ and $v = dy^2$ in Lemma 3.4 to yield $(x^2 + dy^2)^2 = (x^2 - dy^2)^2 + 4dx^2y^2$ or, equivalently, $(x^2 + dy^2)^2 - d(2xy)^2 = (x^2 - dy^2)^2$. So if $x^2 - dy^2 = 1$, then $(x^2 + dy^2)^2 - d(2xy)^2 = 1$, and hence if (x, y) is a solution to $x^2 - dy^2 = 1$ in positive integers, so is $(x^2 + dy^2, 2xy)$. □

The recursion in the proof of Theorem 3.5 generates an infinite sequence of solutions, but the sequence may not contain all solutions. For

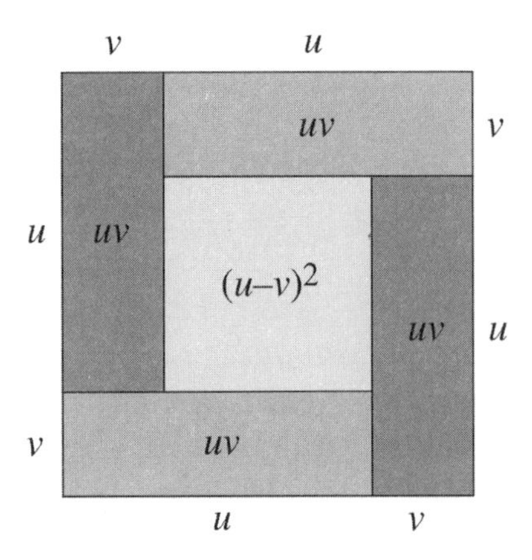

FIGURE 3.12

example, $d = 2$ and $(x, y) = (3, 2)$ generates $(17, 12)$, $(577, 408)$, and so on, omitting half the solutions to $x^2 - 2y^2 = 1$ in Table 3.1.

It is not obvious that $x^2 - dy^2 = 1$ always has a solution in positive integers (when d is positive and non-square). For example, the smallest solution in positive integers to $x^2 - 29y^2 = 1$ is $(x, y) = (9801, 1820)$, and the smallest solution in positive integers to $x^2 - 61y^2 = 1$ is $(x, y) = (1766319049, 226153980)$. Joseph-Louis Lagrange proved in 1766 that $x^2 - dy^2 = 1$ (where d is positive and non-square) has infinitely many solutions in positive integers [Andreescu et al., 2010].

The combinatorial approach in the previous two sections (counting small squares and triangles) to solving Pell equations $x^2 - dy^2 = 1$ becomes difficult for $d \geq 5$. But examination of the results for $d = 2$ and $d = 3$ reveals a pattern:

$$x^2 - 2y^2 = (3x + 4y)^2 - 2(2x + 3y)^2,$$
$$x^2 - 3y^2 = (2x + 3y)^2 - 3(x + 2y)^2.$$

Replacing y by $2y$ in $x^2 - 2y^2 = (3x + 4y)^2 - 2(2x + 3y)^2$ yields

$$x^2 - 8y^2 = (3x + 8y)^2 - 8(x + 3y)^2.$$

The terms to the right of the equal sign each have the form $(ax + dby)^2 - d(bx + ay)^2$, where $a^2 - db^2 = 1$, that is, (a, b) is the smallest solution to $x^2 - dy^2 = 1$ in positive integers from Tables 3.3 and 3.5, and $(a, b) = (3, 1)$ for $d = 8$. This is readily verified for $d = 5$, 6, and 7. For $x^2 - 5y^2 = 1$ we have $(a, b) = (9, 4)$, for $x^2 - 6y^2 = 1$ we have $(a, b) = (5, 2)$, and for $x^2 - 7y^2 = 1$ we have $(a, b) = (8, 3)$ so that

$$x^2 - 5y^2 = (9x + 20y)^2 - 5(4x + 9y)^2,$$
$$x^2 - 6y^2 = (5x + 12y)^2 - 6(2x + 5y)^2,$$
$$x^2 - 7y^2 = (8x + 21y)^2 - 7(3x + 8y)^2,$$

and so on.

We now generate a sequence (x_n, y_n) satisfying $x_n^2 - dy_n^2 = 1$ using the recursions $x_{n+1} = ax_n + dby_n$, $y_{n+1} = bx_n + ay_n$ and starting with an initial solution $(x_0, y_0) = (1, 0)$. Note that $(a, b) = (x_1, y_1)$, so that

$$x^2 - dy^2 = (x_1x + dy_1y)^2 - d(y_1x + x_1y)^2.$$

The recursions $x_{n+1} = x_1x_n + dy_1y_n$ and $y_{n+1} = y_1x_n + x_1y_n$ with $x_1^2 - dy_1^2 = 1$ imply that

$$\begin{aligned} x_{n+1} &= x_1x_n + dy_1(y_1x_{n-1} + x_1y_{n-1}) \\ &= x_1x_n + dy_1^2x_{n-1} + x_1(x_n - x_1x_{n-1}) \\ &= 2x_1x_n + (dy_1^2 - x_1^2)x_{n-1} \\ &= 2x_1x_n - x_{n-1} \end{aligned}$$

and

$$\begin{aligned} y_{n+1} &= y_1(x_1x_{n-1} + dy_1y_{n-1}) + x_1y_n \\ &= x_1(y_n - x_1y_{n-1}) + dy_1^2y_{n-1} + x_1y_n \\ &= 2x_1y_n + (dy_1^2 - x_1^2)y_{n-1} \\ &= 2x_1y_n - y_{n-1}, \end{aligned}$$

i.e., $x_{n+1} = 2x_1x_n - x_{n-1}$ and $y_{n+1} = 2x_1y_n - y_{n-1}$. Evaluating the generating functions $G_x(t) = \sum_{n=0}^{\infty} x_nt^n$ and $G_y(t) = \sum_{n=0}^{\infty} y_nt^n$ for the sequences $\{x_n\}$ and $\{y_n\}$ yields $G_x(t) = (1 - x_1t)/(1 - 2x_1t + t^2)$ and $G_y(t) = y_1t/(1 - 2x_1t + t^2)$, from which we obtain

$$x_n = \frac{1}{2}[(x_1 + y_1\sqrt{d})^n + (x_1 - y_1\sqrt{d})^n]$$

and

$$y_n = \frac{1}{2\sqrt{d}}[(x_1 + y_1\sqrt{d})^n - (x_1 - y_1\sqrt{d})^n]$$

via partial fraction decompositions and geometric series.

Example 3.12. *Square pentagonal numbers.* At first glance it appears that the trivial case $P_1 = 1 = 1^2$ may be the only square pentagonal number. But $P_m = k^2$ is equivalent to $m(3m - 1) = 2k^2$ which, upon completing the square, yields $(6m - 1)^2 - 6(2k)^2 = 1$. Using

TABLE 3.7

n	0	1	2	3	4	5
x_n	1	5	49	485	4801	47525
y_n	0	2	20	198	1960	19402

$x^2 - 6y^2 = (5x + 12y)^2 - 6(2x + 5y)^2$ to solve $x^2 - 6y^2 = 1$ with the initial solution $(x_0, y_0) = (1, 0)$ yields the results in Table 3.7 for n from 0 to 5.

Since $x = 6m - 1$ we require $x_n \equiv -1 \pmod 6$, so that n must be odd. Thus $(x_3, y_3) = (485, 198)$ yields $P_{81} = 9804 = 99^2$, $(x_5, y_5) = (47525, 19402)$ yields $P_{7921} = 94109401 = 9701^2$, and so on. □

3.7. Exercises

3.1 Solve the "coconuts" problem in Example 3.3 with five sailors rather than three.

3.2 Do there exist 1,000,000 consecutive integers each of which contains a repeated prime factor? (This is Problem B4 from the 1955 William Lowell Putnam Competition.)

3.3 Prove that if (x_n, y_n) is a solution to $x^2 - 2y^2 = 1$, then $x_n y_n$ is a multiple of 6.

3.4 Prove that if (x_n, y_n) is a solution to $x^2 - 2y^2 = 1$ or to $x^2 - 2y^2 = -1$, then $(x_n y_n)^2$ is a square triangular number.

3.5 *Ramanujan's house number problem.* The following problem, which appeared in the December 1914 issue of the British magazine *The Strand*, is often mentioned in biographies of the Indian mathematician Srinivasa Ramanujan (1887–1920). The biographies tell us that Ramanujan solved the problem while stirring vegetables in a skillet as a colleague read the problem to him.

> In a certain street there are more than 50 but less than 500 houses in a row, numbered from 1, 2, 3, etc., consecutively. There is a house in the street, the sum of all the house numbers on the left side of which is equal to the sum of the house numbers on its right side. Find the number of this house.

Solve the problem.

3.6 *Powerful numbers.* An integer is called *powerful* if the square of each of its prime divisors also divides it, i.e., no prime divides a powerful number to the first power only. Two pairs of consecutive powerful numbers are $(8, 9) = (2^3, 3^2)$ and $(288, 289) = (2^5 \cdot 3^2, 17^2)$. Prove that infinitely many pairs of consecutive powerful numbers exist.

3.7 Prove that there exist infinitely many integers n such that $n, n+1$, and $n+2$ are each the sum of two squares (of integers). Example: $8 = 2^2 + 2^2, 9 = 3^2 + 0^2, 10 = 3^2 + 1^2$. [This is problem A2 from the 2000 William Lowell Putnam Competition.]

3.8 Observe that $3^2 + 4^2 + 5^2 = 7^2 + 1$. Are there other instances where the sum of the squares of three consecutive integers is one more than a square?

3.9 In Table 1.1 we see that $T_2 = 3T_1$ and $T_9 = 3T_5$, two instances where one triangular number equals three times another. Are there other instances of this phenomenon?

3.10 In Example 1.2 we observed that 225 is both square and octagonal. Show that there are infinitely many square octagonal numbers.

3.11 At the end of Section 1.3 we introduce the star numbers $S_n = 12T_{n-1} + 1 = 6n(n-1) + 1$, and illustrated $S_5 = 121$, i.e., $S_5 = 11^2$. Are there other square star numbers?

3.12 *Almost square triangular numbers.* In the game of pool one is given 16 balls arranged in a square tray, as shown on the left in Figure 3.13. When the white cue ball is removed from the tray, the remaining 15 balls can be placed in a triangle, as shown on the right, since $4^2 - 1 = 15 = T_5$.

FIGURE 3.13

More generally, an *almost square triangular number* is a triangular number that differs from a square by 1, such as $T_2 = 3 = 2^2 - 1$, $T_4 = 10 = 3^2 + 1$, $T_5 = 15 = 4^2 - 1$ (as in the game of pool), and $T_{25} = 325 = 18^2 + 1$. Show that there are infinitely many almost square triangular numbers.

3.13 In $1^3 - 0^3 = 1^2$ and $8^3 - 7^3 = 13^2$ we see that the difference of consecutive cubes is sometimes a square. Are there other instances of this phenomenon?

3.14 Show that there are infinitely many positive integers k such that both $k+1$ and $3k+1$ are perfect squares.

3.15 In Figure 1.3(b) we saw that $T_{k-1} + T_k = k^2$, that is, every pair of consecutive triangular numbers has a square sum. Observe that $T_5 + T_6 + T_7 = 15 + 21 + 28 = 64 = 8^2$. Are there other sets of three consecutive triangular numbers that have a square sum?

3.16 The *inradius* r of a triangle is the radius of its inscribed circle. Prove that the inradius of an almost equilateral Heronian triangle (see Example 3.10) is an integer. [Hint: First show that for an arbitrary triangle, $K = rs$ where K is the area and s is the semiperimeter.]

3.17 An urn contains k balls, of which r are red and $k-r$ are white. Two balls are drawn at random without replacement, and the probability that both are red is $1/2$. What are the possible values of k and r?

CHAPTER 4

Pythagorean Triples

The Hypotenuse has a square on,
which is equal Pythagoras instructed,
to the sum of the squares on the other two sides
If a triangle is cleverly constructed.

Richard Digance

All is number.

Attributed to Pythagoras

The Pythagorean theorem, relating the squares of the lengths of two sides of a right triangle to the square of the length of the third side, may well be the best-known theorem in geometry—maybe in all of mathematics. Of interest in number theory are *Pythagorean triples* (PTs), triples (a, b, c) of positive integers with $a^2 + b^2 = c^2$, so that a, b, and c can be side lengths in a right triangle. Examples include $(3, 4, 5)$, $(6, 8, 10)$, and $(5, 12, 13)$. When a, b, and c are relatively prime—as in $(3, 4, 5)$ and $(5, 12, 13)$—we call the triple a *primitive Pythagorean triple* (PPT). Note that if (x, y, z) is a PT where the greatest common divisor of x, y, and z is d, then $(\frac{x}{d}, \frac{y}{d}, \frac{z}{d})$ is a PPT.

It is easy to show (see Exercise 4.1) that if (a, b, c) is a PPT, then a and b have opposite parity, and c is odd. For convenience in what follows, we will usually write (a, b, c) with a odd and b even. A *Pythagorean triangle* (PΔ) is a triangle whose side lengths are the numbers in a PT. In Figure 4.1(a) we see the PΔ $(3, 4, 5)$ as represented in the *Zhou Bi Suan Jing* (周髀算經), a Chinese text dating from the Zhou dynasty (1046–256 BCE); in Figure 4.1(b) on a Greek postage stamp from 1955; and in Figure 4.1(c) on a Philippine postage stamp from 2001.

Figure 4.2 is an ancient but simple visual proof of the Pythagorean theorem based on the *Zhou Bi Suan Jing*.

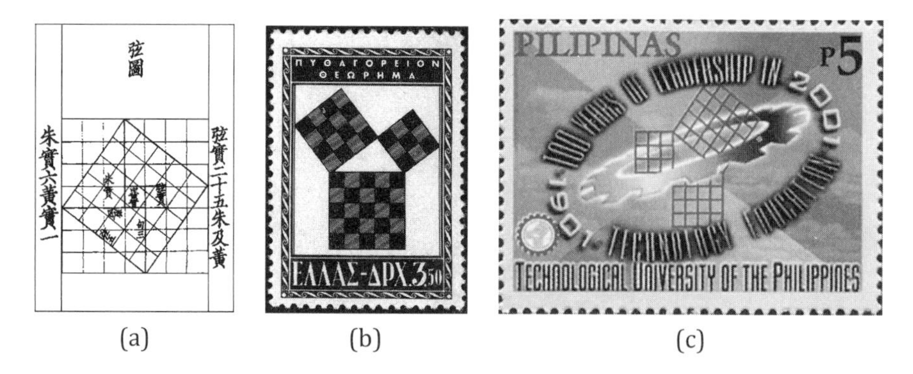

(a) (b) (c)

FIGURE 4.1

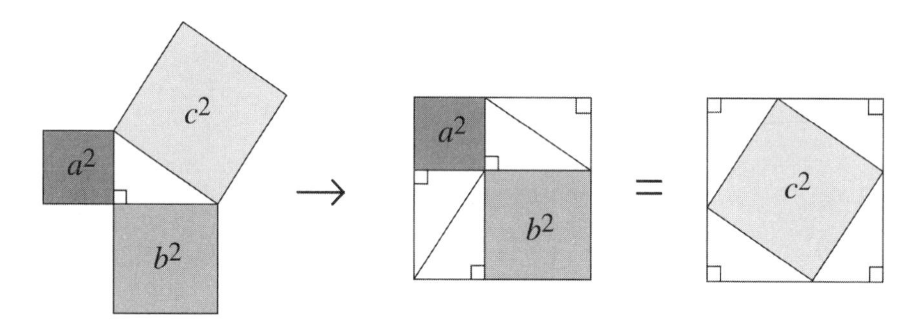

FIGURE 4.2

4.1. Euclid's formula

In Book X of Euclid's *Elements*, Lemma 1 preceding Proposition 29 states: *To find two square numbers such that their sum is also square.* The procedure Euclid employs to accomplish this is equivalent to, in modern notation, the following theorem.

Theorem 4.1. Euclid's formula for PPTs. *Let m and n be relatively prime positive integers with opposite parity with $m > n$. Then $(m^2 - n^2, 2mn, m^2 + n^2)$ is a PPT, and every PPT has this form.*

Figure 4.3 is an illustration of how the pair (m, n) generates the PPT $(m^2 - n^2, 2mn, m^2 + n^2)$.

We call (m, n) the *generator* of $(m^2 - n^2, 2mn, m^2 + n^2)$. A proof that the formula generates PPTs and that every PPT is generated by the formula can be found in most elementary number theory textbooks, e.g., [Marshall et al., 2007].

Euclid's formula can also be illustrated using the double angle formula for the tangent function from elementary trigonometry. If we let

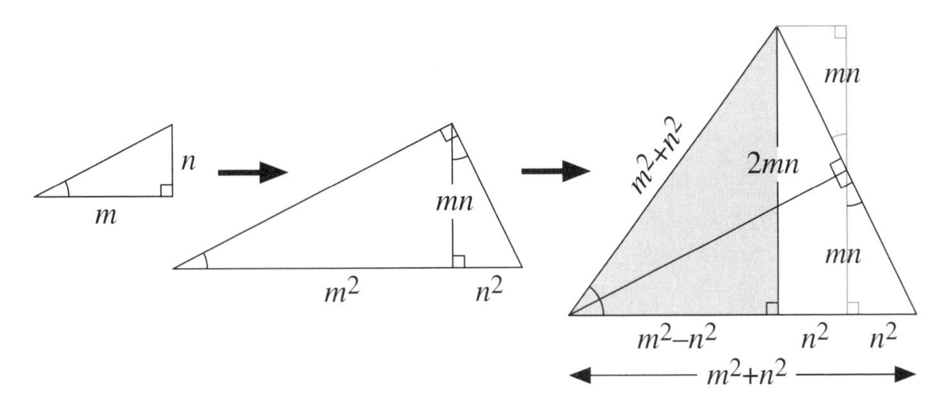

FIGURE 4.3

θ denote the angle marked in the left-most triangle in Figure 4.3, then

$$\tan\theta = \frac{n}{m} \quad \text{implies} \quad \tan 2\theta = \frac{2\tan\theta}{1-\tan^2\theta} = \frac{2n/m}{1-(n/m)^2} = \frac{2mn}{m^2-n^2}.$$

4.2. Pythagorean triples and means of odd squares

There are many ways to compute the average of two positive numbers x and y. Such averages are called *means*, from the French *moyen*, "medium" or "middle." Two of the most common are the *arithmetic mean* $\frac{x+y}{2}$ and the *geometric mean* $\sqrt{xy}$. These means are related by the *arithmetic mean-geometric mean inequality*, which states: *If x and y are positive, then* $\frac{x+y}{2} \geq \sqrt{xy}$. A simple proof of the inequality follows from the observation that the two means are the hypotenuse and one leg of a right triangle, and that the hypotenuse is always the longest side. See Figure 4.4(a).

Now let $x = s^2$ and $y = t^2$, where s and t are relatively prime odd positive integers with $s > t$, as shown in Figure 4.4(b). Then

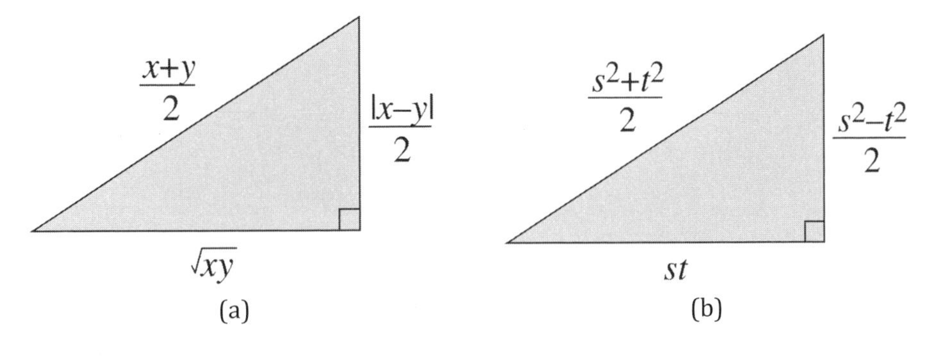

FIGURE 4.4

$\left(st, \frac{s^2-t^2}{2}, \frac{s^2+t^2}{2}\right)$ is a PPT, since this form is equivalent to Euclid's formula in Theorem 4.1 upon setting $s = m + n$ and $t = m - n$. This is the form for PPTs that appears in Euclid's *Elements*. Consequently PPTs arise from computing the geometric and arithmetic means of two relatively prime odd squares (the even leg is half the difference of the squares). For example, the squares 25 and 49 yield the PPT (35,12,37).

4.3. The carpets theorem

The carpets theorem is a little known but quite useful theorem in number theory and combinatorics. We shall use it in several sections in this chapter.

The Carpets Theorem 4.2. *Place two carpets in a room. The area of the floor equals the combined area of the carpets if and only if the area of the overlap equals the area of the uncovered floor.*

See Figure 4.5 for an illustration of two elliptical carpets in a rectangular room.

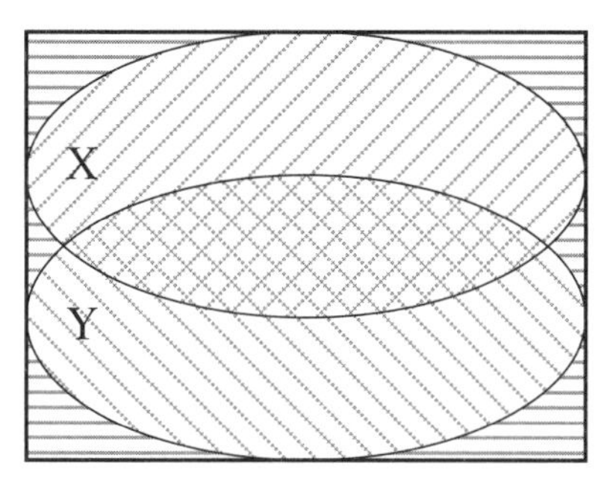

FIGURE 4.5

Proof. Expressing the area of the floor using inclusion-exclusion yields (where A denotes area)

$$\begin{aligned} A(\text{floor}) = A(\text{carpet } X) + A(\text{carpet } Y) \\ - A(\text{overlap}) + A(\text{uncovered floor}), \end{aligned}$$

and thus

$$\begin{aligned} A(\text{floor}) = A(\text{carpet } X) + A(\text{carpet } Y) \\ \Leftrightarrow A(\text{overlap}) = A(\text{uncovered floor}). \end{aligned}$$ □

The shapes of the room and the carpets are arbitrary, and the result holds for more than two carpets as long as at most two carpets overlap simultaneously. In the next chapter we use the carpets theorem to give proofs that $\sqrt{2}$ and $\sqrt{3}$ are irrational.

4.4. Pythagorean triples and the factors of even squares

Several of the results we are about to derive for PTs and PPTs follow from a simple two-part lemma.

Lemma 4.3. *For positive x, y, and z we have*

$$\text{(i) } (x+y+z)^2 = (x+z)^2 + (y+z)^2 \Leftrightarrow z^2 = 2xy, \tag{4.1}$$

$$\text{(ii) } (x+y+z)^2 = 2(x+z)(y+z) \Leftrightarrow x^2 + y^2 = z^2. \tag{4.2}$$

Proof. In Figure 4.6 we see two possible placements of overlapping carpets in a square room.

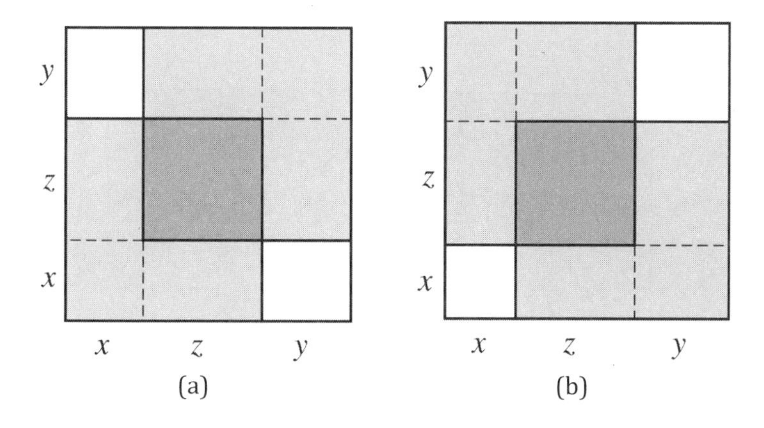

FIGURE 4.6

Applying the carpets theorem to Figure 4.6(a) yields (4.1), and applying the carpets theorem to Figure 4.6(b) yields (4.2). □

We now have a second method to generate PTs and PPTs in the following theorem.

Theorem 4.4. *There exists a one-to-one correspondence between PTs and factorizations of even squares of the form $z^2 = 2xy$.*

Proof ([Gomez, 2005]). Let (a, b, c) be a PT, and set $x = c - b$, $y = c - a$, and $z = a + b - c$. Then $x + y + z = c$, $x + z = a$, and $y + z = b$, and (4.1) yields $c^2 = a^2 + b^2$ if and only if $z^2 = 2xy$. Note that the PT $(a, b, c) = (x + z, y + z, x + y + z)$ is primitive if and only if x and y are relatively prime. □

Here are some examples of the use of Theorem 4.4.

Example 4.1. There are three factorizations of 36 in the form $z^2 = 2xy$, yielding three PTs: $6^2 = 2 \cdot 1 \cdot 18$ yields the PPT (7,24,25); $6^2 = 2 \cdot 2 \cdot 9$ yields the PPT (8,15,17); and $6^2 = 2 \cdot 3 \cdot 6$ yields the PT (9,12,15). Note that when x and y are relatively prime, the PT is primitive. □

Example 4.2. Since $(2k)^2 = 2 \cdot 1 \cdot 2k^2$ for every positive integer k, we have the PPT $(2k+1, 2k^2+2k, 2k^2+2k+1)$, PPTs of the form $(a, b, b+1)$, i.e., the hypotenuse c is one more than the even leg b, and $a^2 = 2b+1$. The generators are consecutive integers, i.e., $(m, n) = (k+1, k)$. Furthermore, every odd number greater than 1 appears as a leg of some PPT, and if c and d are *consecutive* positive integers, then $(c+d, 2cd, c^2+d^2)$ is a PPT (the one above when $c = k$ and $d = k+1$). □

In Figure 4.7(a) we illustrate $a^2+b^2=(b+1)^2$ for the PPT $(a, b, b+1)$ with $a^2=2b+1$ (the area of the darker gray L-shaped region).

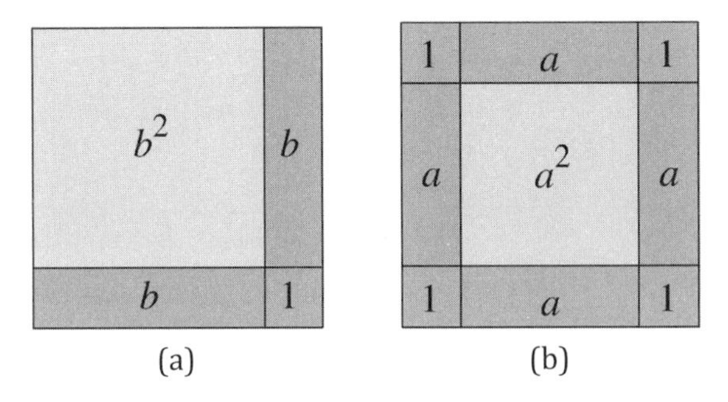

FIGURE 4.7

Example 4.3. Since $(2k)^2 = 2 \cdot 2 \cdot k^2$ for every *odd* integer k (so that 2 and k^2 are relatively prime) we have the PPT $(k^2+2k, 2k+2, k^2+2k+2)$, PPTs of the form $(a, b, a+2)$, i.e., the hypotenuse c is two more than the odd leg a, and $b^2 = 4a+4$. The generators are $(m, n) = (k+1, 1)$ with k odd. □

In Figure 4.7(b) we illustrate $a^2+b^2 = (a+2)^2$ for the PPT $(a, b, a+2)$ with $b^2 = 4a+4$ (the area of the dark gray region bordering the light gray square).

Example 4.4. The factorization $n^2 = 2k(k+1)$ is equivalent to $(n+2k+1)^2 = (n+k)^2 + (n+k+1)^2$. Note that the legs $n+k$ and $n+k+1$ are consecutive integers (which are relatively prime). In this case we refer to the PPT as *almost isosceles*. For example, $2^2 = 2 \cdot 1 \cdot 2$ yields the PPT $(3, 4, 5)$, and $12^2 = 2 \cdot 8 \cdot 9$ yields the PPT $(21, 20, 29)$. Are there other almost isosceles PPTs? If so, how many? □

4.5. Almost isosceles primitive Pythagorean triples

To answer the questions at the end of Example 4.4, we have

Theorem 4.5. *Infinitely many almost isosceles PPTs exist.*

We present two proofs, the first using the carpets theorem, the second via one of the Pell equations we solved in Chapter 3.

Proof 1. Let $(a, a+1, c)$ or $(a+1, a, c)$ be an almost isosceles PPT. Place carpets with areas $(3a+2c+1)^2$ and $(3a+2c+2)^2$ in a room with area $(4a+3c+2)^2$, as shown in Figure 4.8(a).

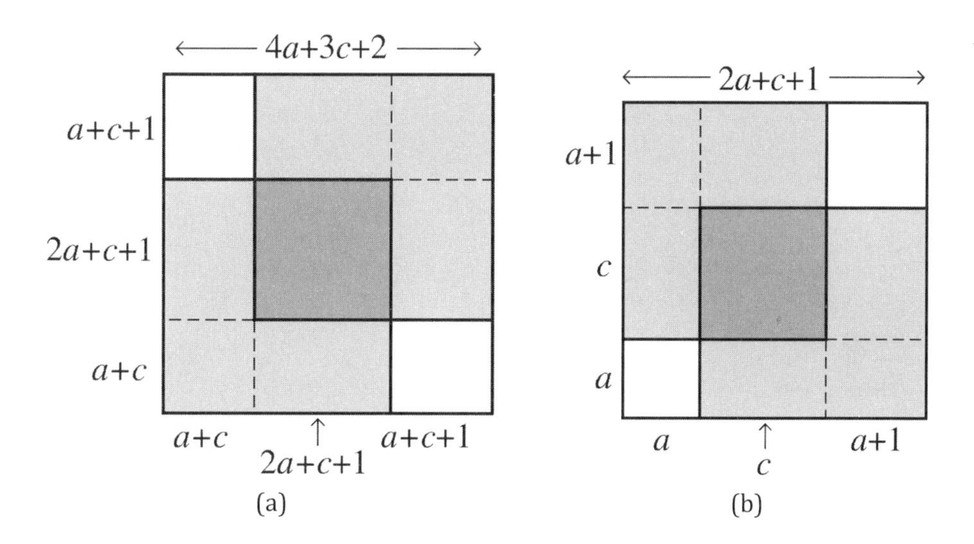

FIGURE 4.8

The carpets theorem yields

$$(4a+3c+2)^2 = (3a+2c+1)^2 + (3a+2c+2)^2$$
$$\Leftrightarrow (2a+c+1)^2 = 2(a+c)(a+c+1).$$

Now place two carpets each with area $(a+c)(a+c+1)$ in a room with area $(2a+c+1)^2$, as shown in Figure 4.8(b). The carpets theorem yields

$$(2a+c+1)^2 = 2(a+c)(a+c+1) \Leftrightarrow a^2 + (a+1)^2 = c^2.$$

Combining the two equivalences yields

$$(4.3) \quad a^2 + (a+1)^2 = c^2$$
$$\Leftrightarrow = (3a + 2c + 1)^2 + (3a + 2c + 2)^2 = (4a + 3c + 2)^2,$$

i.e., each almost isosceles PPT yields a larger one. This generates the sequence (3, 4, 5), (21, 20, 29), (119, 120, 169), (697, 696, 985), etc. □

Proof 2. Let $(a, a+1, c)$ or $(a+1, a, c)$ be an almost isosceles PPT. In Figure 4.9 we show that two copies of this triangle (in light gray), an isosceles right triangle with legs c (in white), and a right triangle with legs 1 and $2a + 1$ and hypotenuse $c\sqrt{2}$ (in darker gray) partition a $(2a + 1) \times (a + 1)$ rectangle.

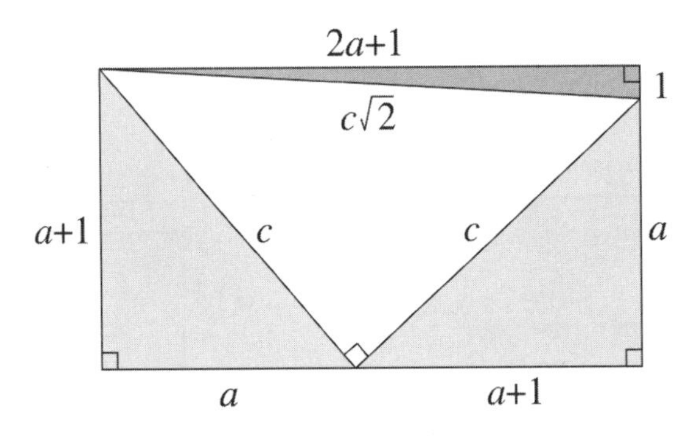

FIGURE 4.9

Hence $(2a + 1)^2 + 1 = 2c^2$, equivalent to $(2a + 1)^2 - 2c^2 = -1$, the second Pell equation $x^2 - 2y^2 = -1$ we solved in Section 3.4 with $x = 2a+1$ and $y = c$. Each solution (x_n, y_n) for $n \geq 1$ in Table 3.4 has x odd and $y \geq 5$, and hence leads to an almost isosceles PPT. For example, $(x_1, y_1) = (7, 5)$ leads to the PPT (3,4,5), $(x_2, y_2) = (41, 29)$ leads to the PPT (21,20,29), $(x_3, y_3) = (239, 169)$ leads to the PPT (119,120,169), etc. □

The next two corollaries exploit relationships between almost isosceles PPTs and triangular numbers.

Corollary 4.6. *There exist infinitely many PTs in which the legs are consecutive triangular numbers.*

Proof. In Figure 4.10 we illustrate with Pythagorean triangles, expanding the $(a, a+1, c)$ triangle by a factor of $2a + 1$ to form a Pythagorean triangle with legs of length T_{2a} and T_{2a+1}. □

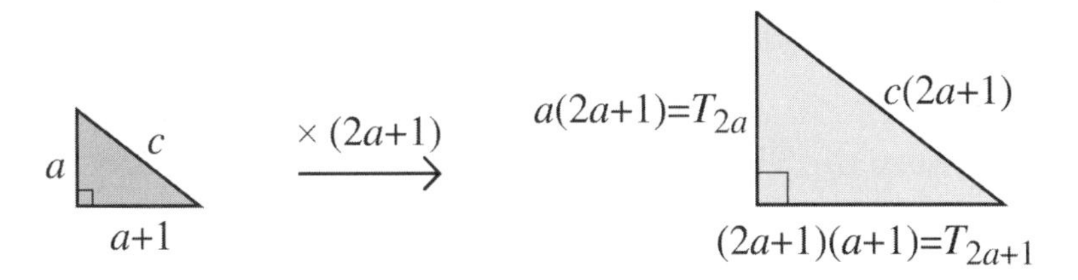

FIGURE 4.10

We note that it is possible to find a PT whose three sides are triangular numbers, e.g., $(T_{132}, T_{143}, T_{164})$. However, it is unknown whether the number of such PTs is finite or infinite.

In Chapters 1, 2, and 3 we proved that infinitely many square triangular numbers exist. Here is yet another proof using almost isosceles PPTs.

Corollary 4.7. *Infinitely many square triangular numbers exist.*

Proof. In the first proof of Theorem 4.5 we showed that $a^2 + (a+1)^2 = c^2$ is equivalent to $(2a+c+1)^2 = 2(a+c)(a+c+1) = 4T_{a+c}$. Since $c+1$ is even, we have $T_{a+c} = \left(a + \frac{c+1}{2}\right)^2$. Thus there is a one-to-one correspondence between almost isosceles PPTs and square triangular numbers. Since there are infinitely many almost isosceles PPTs, there are also infinitely many square triangular numbers. □

Thus $3^2 + 4^2 = 5^2$ implies $T_8 = 6^2$, $20^2 + 21^2 = 29^2$implies $T_{49} = 35^2$, $119^2 + 120^2 = 169^2$ implies $T_{288} = 204^2$, etc.

We conclude this section by showing how almost isosceles PPTs solve the Pell equations $x^2 - 2y^2 = \pm 1$ encountered in Section 3.4.

Theorem 4.8. *If (m, n) is the generator of an almost isosceles PPT (a, b, c), where a and b differ by* 1*, then $(x, y) = (m-n, n)$ is a solution to either $x^2 - 2y^2 = +1$ or $x^2 - 2y^2 = -1$.*

Proof. If $a - b = \pm 1$, then $\left(m^2 - n^2\right) - 2mn = \pm 1$ or, equivalently, $(m-n)^2 - 2n^2 = \pm 1$. □

It remains to show how to find the generators of the almost isosceles PPTs. Setting $2a + 1 = a + (a+1) = a + b$ in the hypotenuse term $4a + 3c + 2$ in (4.3) yields

$$4a + 3c + 2 = 2a + 2b + 3c = 5m^2 + 4mn + n^2 = (2m+n)^2 + m^2,$$

so if (m, n) generates one almost isosceles PPT, $(2m + n, m)$ generates the next one. Hence the sequence of generators is (m, n) = (2,1), (5,2), (12,5), (29,12), and so on, yielding the sequence of solutions (x, y) = (1,1), (3,2), (7,5), (17,12), etc., to $x^2 - 2y^2 = \pm 1$ (the signs alternate, beginning with $-$).

4.6. A Pythagorean triple tree

There are several ways to arrange all the PPTs into a *trinary tree*, a structure where the PPT $(3,4,5)$ "branches" into three different PPTs, and each of those branches into three different PPTs, and so on. Here is one example of such a tree [Price, 2011] , showing the first four "generations":

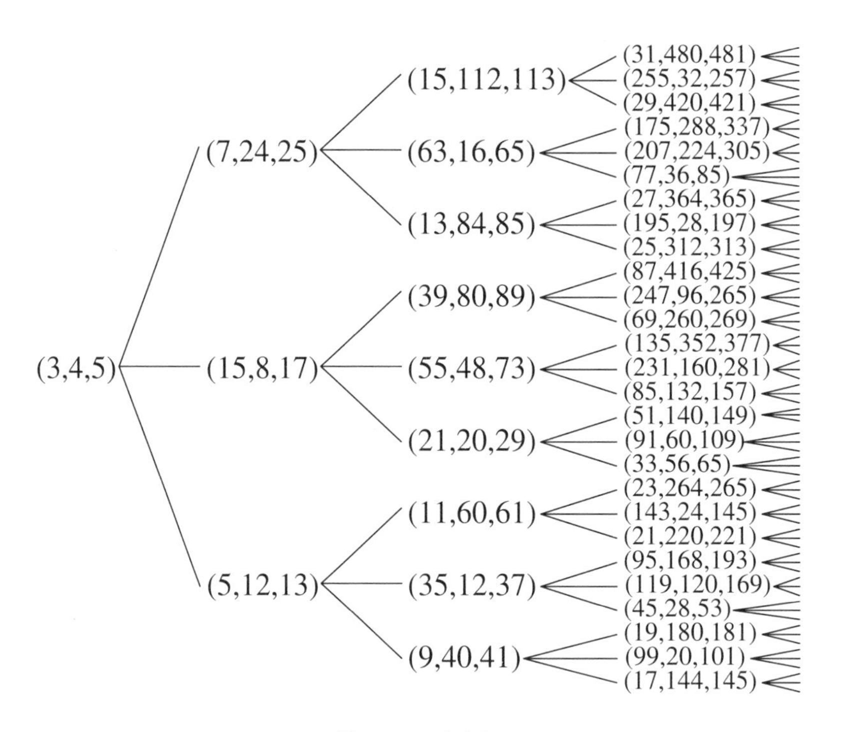

FIGURE 4.11

The tree in Figure 4.11 is generated by three linear transformations of a triple (a,b,c), one,

$$U\,(a,b,c) = (2a - b + c, 2a + 2b + 2c, 2a + b + 3c)$$

for each "upward" sloping ↗ branch, another,

$$A\,(a,b,c) = (2a + b + c, 2a - 2b + 2c, 2a - b + 3c)$$

for each "across" → branch, and a third,

$$D\,(a,b,c) = (2a + b - c, -2a + 2b + 2c, -2a + b + 3c)$$

for each "downward" sloping ↘ branch. To show that U, A, and D generate PTs, we need to show that $U\,(a,b,c)$, $A\,(a,b,c)$, and $D\,(a,b,c)$ are PTs whenever (a,b,c) is a PT. The proof employs the carpets theorem.

Theorem 4.9. *Let a, b, c be positive numbers. Then*

(4.4)
$$c^2 = a^2 + b^2 \Leftrightarrow (2a+b+3c)^2 = (2a-b+c)^2 + (2a+2b+2c)^2,$$

(4.5)
$$c^2 = a^2 + b^2 \Leftrightarrow (2a-b+3c)^2 = (2a+b+c)^2 + (2a-2b+2c)^2,$$

and

(4.6)
$$c^2 = a^2 + b^2 \Leftrightarrow (-2a+b+3c)^2 = (2a+b-c)^2 + (-2a+2b+2c)^2.$$

Consequently $U(a,b,c)$, $A(a,b,c)$, and $D(a,b,c)$ are PTs if and only if (a,b,c) is a PT.

Proof. (i) Place carpets with areas $(2a-b+c)^2$ and $(2a+2b+2c)^2$ in a room with area $(2a+b+3c)^2$, as shown in Figure 4.12(a).

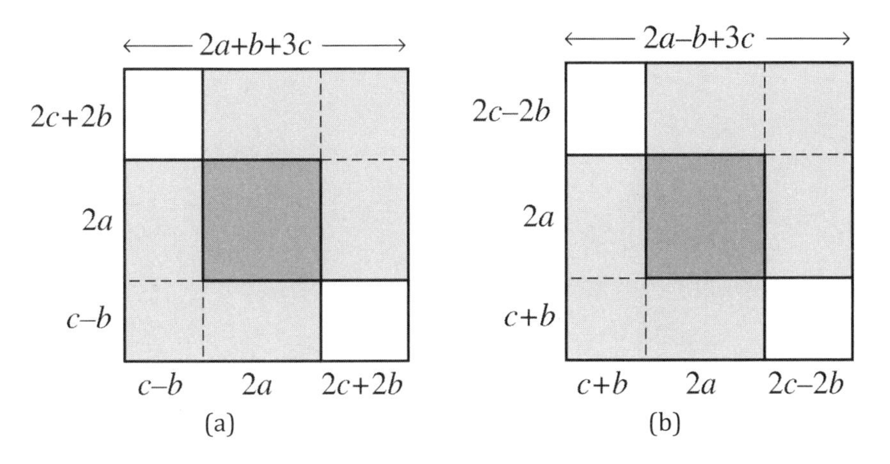

FIGURE 4.12

The carpets theorem yields

$$(2a+b+3c)^2 = (2a-b+c)^2 + (2a+2b+2c)^2$$
$$\Leftrightarrow (2a)^2 = 2(2c+2b)(c-b).$$

But $(2a)^2 = 2(2c+2b)(c-b)$ is equivalent to $a^2 = c^2 - b^2$, from which (4.4) follows. Thus $U(a,b,c)$ is a PT if and only if (a,b,c) is a PT.

(ii) Place carpets with areas $(2a+b+c)^2$ and $(2a-2b+2c)^2$ in a room with area $(2a-b+3c)^2$, as shown in Figure 4.12(b). The carpets theorem yields

$$(2a - b + 3c)^2 = (2a + b + c)^2 + (2a - 2b + 2c)^2$$
$$\Leftrightarrow (2a)^2 = 2(2c - 2b)(c + b).$$

But $(2a)^2 = 2(2c - 2b)(c + b)$ is equivalent to $a^2 = c^2 - b^2$, from which (4.5) follows. Thus $A(a, b, c)$ is a PT if and only if (a, b, c) is a PT.

(iii) Place carpets with areas $(2a + b - c)^2$ and $(-2a + 2b + 2c)^2$ in a room with area $(-2a + b + 3c)^2$, as shown in Figure 4.13(a).

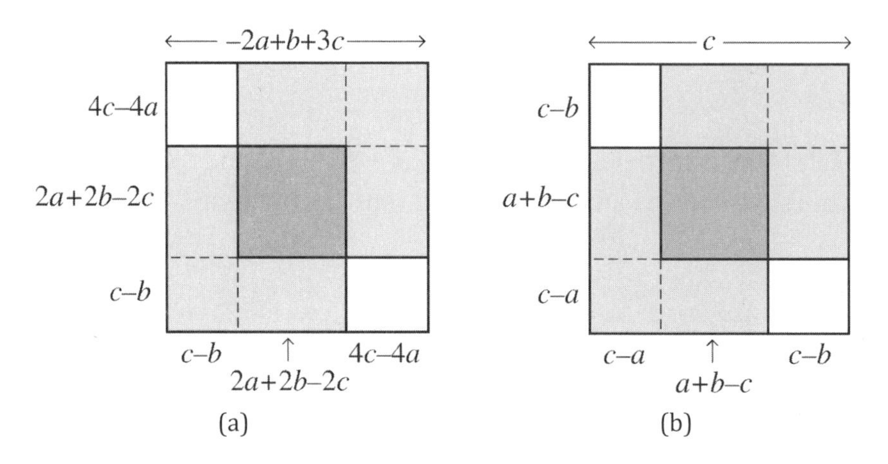

FIGURE 4.13

The carpets theorem yields

$$(-2a + b + 3c)^2 = (2a + b - c)^2 + (-2a + 2b + 2c)^2$$
$$\Leftrightarrow (2a + 2b - 2c)^2 = 2(c - b)(4c - 4a)$$

or, equivalently,

$$(-2a + b + 3c)^2 = (2a + b - c)^2 + (-2a + 2b + 2c)^2$$
$$\Leftrightarrow (a + b - c)^2 = 2(c - b)(c - a).$$

Now place carpets with areas a^2 and b^2 in a room with area c^2, as shown in Figure 4.13(b). The carpets theorem yields

$$c^2 = a^2 + b^2 \Leftrightarrow (a + b - c)^2 = 2(c - b)(c - a),$$

and combining the two equivalences yields (4.6). Thus $D(a, b, c)$ is a PT if and only if (a, b, c) is a PT. □

See [Price, 2011] for proofs that only PPTs appear in the tree and that every PPT appears exactly once. For another trinary tree of Pythagorean triples, see [Brousseau, 1972].

4.7. Primitive Pythagorean triples with square sides

Examination of the tree in Figure 4.11 shows that some PPTs have an odd square leg, e.g., (9,40,41) and (25,312,313); some have an even square leg, e.g., (63,16,65) and (77,36,85); and some have a square hypotenuse, e.g., (7,24,25) and (119,120,169). We now show that there are infinitely many PPTs of each type, using Lemma 3.4.

Theorem 4.10. *There are* (a) *infinitely many PPTs with an odd square leg;* (b) *infinitely many PPTs with an even square leg; and* (c) *infinitely many PPTs with a square hypotenuse.*

Proof. (a) Let (a,b,c) be a PPT (with a odd), and set $(u,v)=(c^2,b^2)$ in Lemma 3.4, so that $\left(c^2-b^2\right)^2+4b^2c^2=\left(c^2+b^2\right)^2$. Hence $(a^2,2bc,$ $b^2+c^2)$ is a PPT with an odd square leg. The PT is primitive since its generator is $(m,n)=(c,b)$.

(b) Let k be a positive integer, and set $(u,v)=(4k^4,1)$ in Lemma 3.4, so that $(4k^4-1)^2+16k^4=(4k^4+1)^2$. Hence $(4k^4-1,4k^2,4k^4+1)$ is a PPT (since its generator is $(m,n)=(2k^2,1)$). For another family of PPTs with an even square leg, set $(u,v)=((2k+1)^4,2)$ in Lemma 3.4, yielding the PPT $\left((2k+1)^4-4,4(2k+1)^2,(2k+1)^4+4\right)$.

(c) Let (a,b,c) be a PPT, and set $(u,v)=(b^2,a^2)$ in Lemma 3.4, so that $\left(b^2-a^2\right)^2+4a^2b^2=(b^2+a^2)^2$. Hence $(\left|b^2-a^2\right|,2ab,c^2)$ is a PPT with a square hypotenuse since its generator is $(m,n)=(b,a)$ or (a,b). ☐

Corollary 4.11. *There are infinitely many solutions to the Diophantine equations* $x^4+y^2=z^2$ *(e.g.,* $(x,y,z)=(2,3,5)$ *or* $(3,40,41)$*) and* $x^2+y^2=z^4$ *(e.g.,* $(x,y,z)=(7,24,5)$*).*

See Exercise 4.14 for solutions to the Diophantine equation $x^2+y^2=z^3$.

Are there PTs with two or three square sides? The answer is no; see [Sierpiński, 2003] for a proof.

4.8. Pythagorean primes and triangular numbers

If the hypotenuse c in a PPT (a,b,c) is prime, then it is called a *Pythagorean prime*. The Pythagorean primes less than 100 are 5, 13, 17, 29, 37, 41, 53, 61, 73, 89, and 97. It follows from Euclid's formula that a Pythagorean prime p equals m^2+n^2, where m and n have opposite parity, hence $p=4N+1$ for some N. We now show that N is the sum of two triangular numbers (one of which may be $T_0=0$). Although we won't prove it here, if the hypotenuse c in a PPT (a,b,c) is composite, it must be a product of Pythagorean primes.

Theorem 4.12. *Let $p = 4N + 1 = (2u)^2 + (2v + 1)^2$, $u \geq 1$, $v \geq 0$, be a Pythagorean prime. Then N is the sum of two triangular numbers, i.e.,*

$$(1)\ N = T_{u+v} + T_{u-v-1} \text{ if } u > v, \text{ and } (2)\ N = T_{u+v} + T_{v-u} \text{ if } u \leq v.$$

Proof. First note that $N = u^2 + v(v + 1)$. Figure 4.14 shows that $u > v$ implies $N = T_{u+v} + T_{u-v-1}$, and Figure 4.15 shows that $u \leq v$ implies $N = T_{u+v} + T_{v-u}$. □

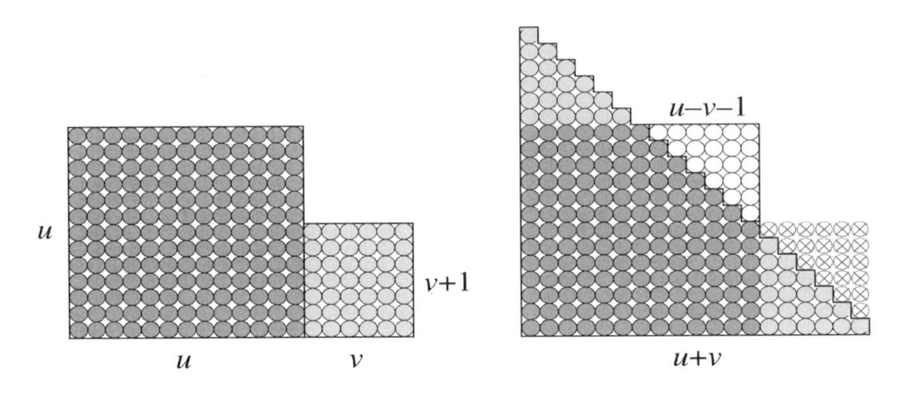

FIGURE 4.14

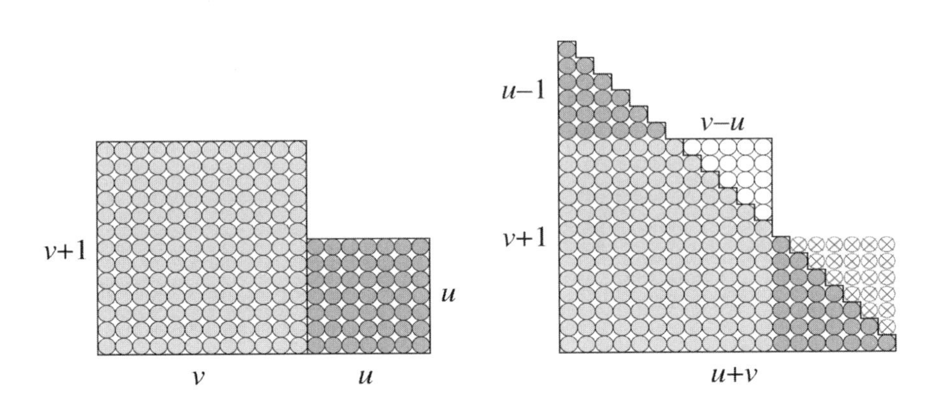

FIGURE 4.15

A converse of Theorem 4.12 also holds.

Theorem 4.13. *If N is the sum of two triangular numbers, then $4N + 1$ is the sum of two integer squares with opposite parity.*

Proof. Let $N = T_u + T_v$ with $u \geq v \geq 0$. Simple algebra establishes $4N+1 = 4(T_u + T_v)+1 = (u + v + 1)^2 + (u - v)^2$, which we illustrate in Figure 4.16 for $(u, v) = (8, 3)$. Observe that $u + v + 1$ and $u - v$ have opposite parity. □

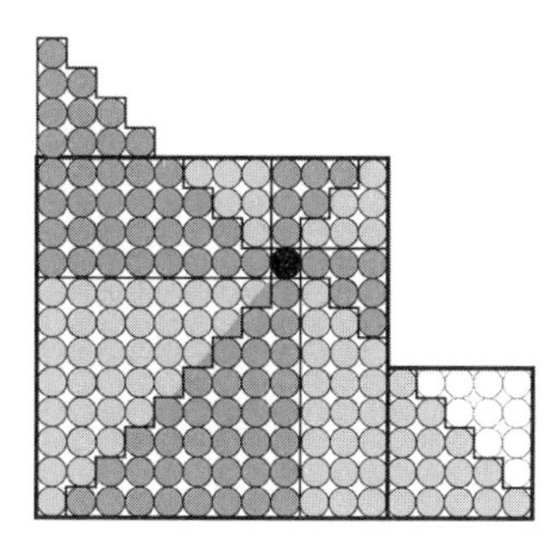

FIGURE 4.16

Note that $4N + 1$ need not be a Pythagorean prime in Theorem 4.13. For example, when $N = T_3 + T_4 = 16$, $4N + 1 = 65$. However, it can be shown that each composite hypotenuse in a PPT is a sum of two squares with opposite parity, and that Theorem 4.12 also holds in this case.

In the following corollary we employ the preceding theorems to generate PPTs using triangular numbers. The proof follows from using $m = u + v + 1$ and $n = u - v$ in Euclid's formula.

Corollary 4.14. *If* $u + v + 1$ *and* $u - v$ *with* $u > v$ *are relatively prime positive integers, then* $((2u + 1)(2v + 1), 4(T_u - T_v), 4(T_u + T_v) + 1)$ *is a PPT.*

Example 4.5. Setting $v = 0$ in the above corollary yields the PPT $(2u + 1, 2u^2 + 2u, 2u^2 + 2u + 1)$, a member of the family of PPTs of the form $(a, b, b + 1)$ from Example 4.2. Setting $v = u - 1$ in the corollary yields the PPT $(4u^2 - 1, 4u, 4u^2 + 1)$, a member of the family of PPTs of the form $(a, b, a + 2)$ from Example 4.3. □

The triangular numbers T_u and T_v also appear in the odd leg $a = (2u + 1)(2v + 1)$ of the PPT in Corollary 4.14, since Lemma 2.4 implies that $a^2 = (8T_u + 1)(8T_v + 1)$.

4.9. Divisibility properties

A cursory examination of the triples in the tree in Figure 4.11 leads one to suspect that in every PT one side is divisible by 3, one side is divisible by 4, and one side is divisible by 5. We now use the result in Corollary 4.14 to show that this is indeed true.

Example 4.6. The divisibility properties mentioned in the preceding paragraph may involve all three sides of a PPT, such as in (3,4,5) and (63,16,65); involve only two sides, such as in (15,8,17) and (7,24,25); or just one side, such as in (11,60,61) and (119,120,169). □

Theorem 4.15. *In every Pythagorean triple, one leg is divisible by* 3*, one leg is divisible by* 4*, and one of the three sides is divisible by* 5*.*

Proof. It suffices to prove the theorem for PPTs, since every PT is either a PPT or an integer multiple of one. Consider the PPT (a, b, c) with generator (m, n) with $m > n$, m and n relatively prime and of opposite parity, and set $u = (m + n - 1)/2$ and $v = (m - n - 1)/2$. Then $m = u + v + 1$ and $n = u - v$ so that

$$(a, b, c) = ((2u + 1)(2v + 1), 4(T_u - T_v), 4(T_u + T_v) + 1)$$

as in Corollary 4.14. Clearly $4|b$. If $u \equiv 1 \pmod 3$, then $2u + 1 \equiv 0 \pmod 3$ so that $3|a$. Similarly if $v \equiv 1 \pmod 3$, then $2v + 1 \equiv 0 \pmod 3$ so that $3|a$. If both $u \not\equiv 1 \pmod 3$ and $v \not\equiv 1 \pmod 3$, then as shown in Theorem 2.2, both T_u and T_v are congruent to 0 (mod 3), and hence $3|b$.

If $u \equiv 2 \pmod 5$, then $2u + 1 \equiv 0 \pmod 5$ so that $5|a$. Similarly if $v \equiv 2 \pmod 5$, then $2v + 1 \equiv 0 \pmod 5$ so that $5|a$. If both $u \not\equiv 2 \pmod 5$ and $v \not\equiv 2 \pmod 5$, then as shown in Theorem 2.3, each of T_u and T_v is congruent to 0 or to 1 (mod 5). If both T_u and T_v are congruent to 0 or both are congruent to 1, then $4\,(T_u - T_v) \equiv 0 \pmod 5$ so that $5|b$. If one of T_u or T_v is congruent to 0 and the other is congruent to 1, then $4\,(T_u + T_v) + 1 \equiv 0 \pmod 5$ so that $5|c$, which completes the proof. □

4.10. Pythagorean triangles

We begin by considering the *inradius* of a triangle, the radius of its *incircle* (the triangle's inscribed circle), as illustrated in Figure 4.17 for a right triangle, and show that the inradius r is a linear function of the lengths of the three sides.

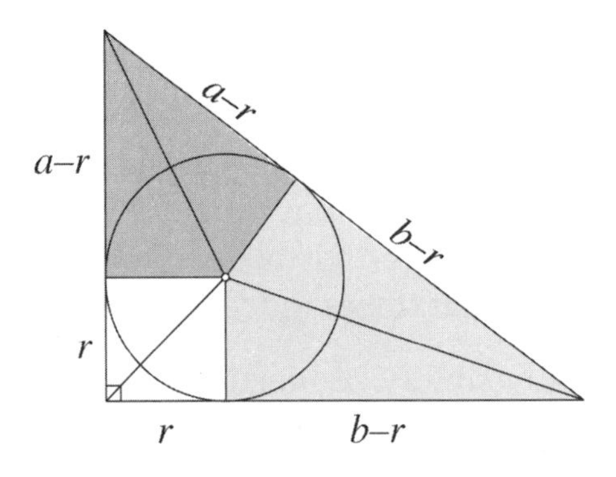

FIGURE 4.17

Lemma 4.16. *The inradius r of a right triangle with legs a and b and hypotenuse c is $r = \frac{1}{2}(a+b-c)$.*

Proof. See Figure 4.17, which illustrates how the angle bisectors of the triangle and radii of the incircle partition the triangle into six smaller triangles, which are congruent in pairs. Consequently the hypotenuse c equals $a+b-2r$, from which the result follows. □

Theorem 4.17. *The inradius of a Pythagorean triangle is an integer.*

Proof. Consider the P△ (ka, kb, kc), where (a,b,c) is a PPT and k a positive integer. Then its inradius $r = \frac{k}{2}(a+b-c)$ is an integer since $a+b-c$ is even. If the generator of (a,b,c) is (m,n), then the inradius r of the P△ (ka, kb, kc) is $kn(m-n)$. □

Example 4.7. In the proof of Theorem 4.4 we encountered PTs of the form $(x+z, y+z, x+y+z)$ which arise from factorizations of even squares of the form $z^2 = 2xy$. Here the inradius of a P△ is $r = z/2$ (an integer since z is even). So it is a simple matter to find all P△'s with a given inradius: Factor the square of the diameter $z = 2r$, e.g., factoring 6^2 yields three P△'s with $r = 3$: (7,24,25), (15,8,17), and (9,12,15). See Figure 4.18. □

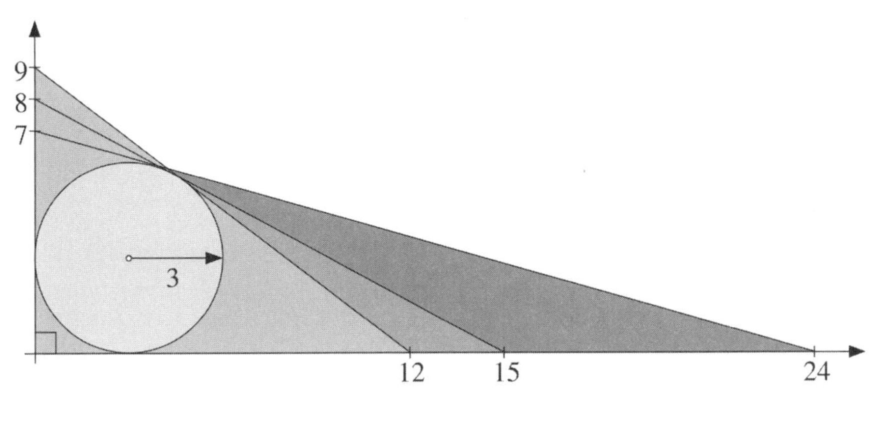

FIGURE 4.18

In addition to an incircle that is tangent to each of the three sides, every triangle possesses three *excircles*, each tangent to one side and to extensions of the other two sides. In Figure 4.19(a) we see the three excircles, with their *excenters* I_a, I_b, I_c, and *exradii* r_a, r_b, r_c, the subscript indicating the triangle's side of tangency to the excircle. We have illustrated the circles for a right triangle, although all triangles possess the three circles, centers, and radii.

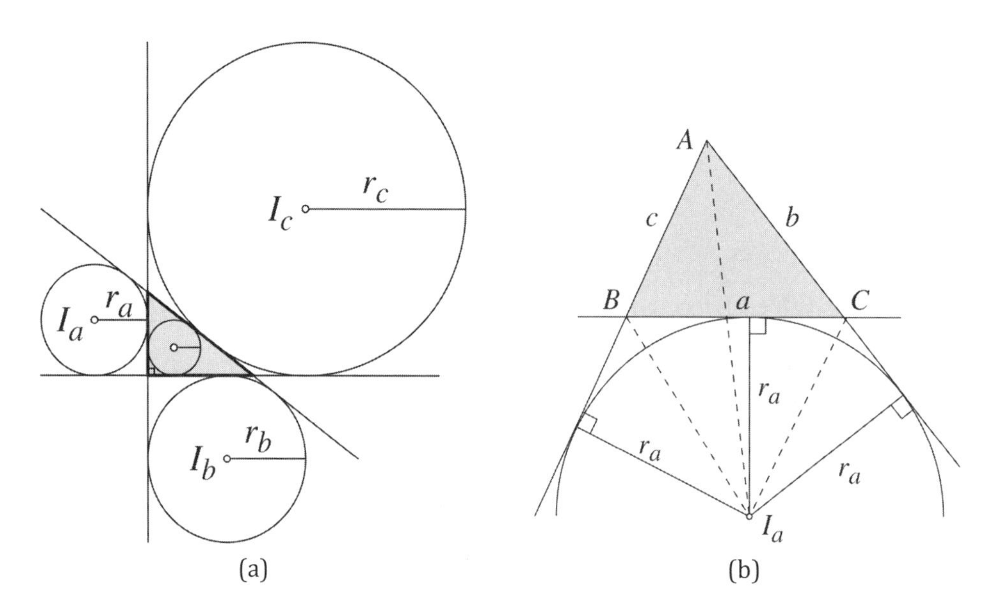

FIGURE 4.19

Theorem 4.18. *The three exradii of a Pythagorean triangle are integers.*

Proof. In Figure 4.19(b) we see a general triangle ABC with a portion of its excircle with excenter I_a and exradius r_a. The area K of ABC is equal to the sum of the areas of triangles AI_aC and AI_aB minus the area of triangle BI_aC, hence

$$K = \frac{1}{2}br_a + \frac{1}{2}cr_a - \frac{1}{2}ar_a = r_a\frac{b+c-a}{2},$$

and hence $r_a = 2K/(b+c-a)$. Similarly $r_b = 2K/(a-b+c)$ and $r_c = 2K/(a+b-c)$.

Now consider a primitive P△ (a,b,c) with generator (m,n), perimeter $P = a+b+c = 2m(m+n)$, and area $K = ab/2 = mn(m^2-n^2)$. Then

$$r_a = \frac{ab}{b+c-a} = m\,(m-n)\,.$$

Similarly $r_b = \frac{ab}{a-b+c} = n\,(m+n)$ and $r_c = \frac{ab}{a+b-c} = m\,(m+n)$, hence all three exradii are integers. In a right triangle the exradius r_c of the excircle on the hypotenuse equals the *semiperimeter* $s = P/2$ of (a,b,c). It now follows that every P△ has integer exradii. □

Example 4.8. The P△ in Figure 4.19(a) is the (3,4,5) triangle with generator $(m,n) = (2,1)$ and thus $(r, r_a, r_b, r_c) = (1,2,3,6)$. □

See Exercise 4.10 for some remarkable relationships among the inradius r, the three exradii, the perimeter, and the area of a P△.

4.11. Pythagorean runs

The PPT (3,4,5) is the only triple where the sum of two consecutive squares is the next square. But what about sums of three consecutive squares? Four consecutive squares? Observe:

$$3^2 + 4^2 = 5^2;$$

$$10^2 + 11^2 + 12^2 = 13^2 + 14^2;$$

$$21^2 + 22^2 + 23^2 + 24^2 = 25^2 + 26^2 + 27^2; \text{ etc.}$$

What's the pattern? A little reflection yields the observation that each number squared immediately to the left of the equals sign is four times a triangular number, i.e., $4 = 4(1)$, $12 = 4(1 + 2)$, $24 = 4(1 + 2 + 3)$, etc., which leads to the identity

$$(4.7) \qquad (4T_n - n)^2 + \cdots + (4T_n)^2 = (4T_n + 1)^2 + \cdots + (4T_n + n)^2.$$

The identity can be easily verified by mathematical induction, but Figures 4.20 and 4.21 (for the case $n = 3$) may better explain why the relationship holds [Boardman, 2000]. In Figure 4.20 we use the fact that $4T_3 = 4 \cdot 1 + 4 \cdot 2 + 4 \cdot 3$ to dissect $4T_3$ into three sets of vertical strips.

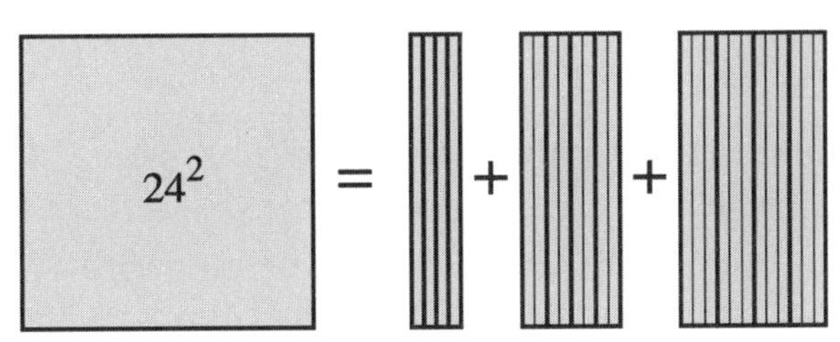

FIGURE 4.20

In Figure 4.21 we use the strips to create "borders" around the remaining three squares on the left side of the equals sign in (4.7) to create the three squares on the right side of the equals sign, yielding $21^2 + 22^2 + 23^2 + 24^2 = 25^2 + 26^2 + 27^2$.

4.12. Sums of two squares

The problem of representing an integer as the sum of two squares has a long history, dating back to at least the time of Pythagoras. In Section 4.8 we discussed Pythagorean primes—primes $p \equiv 1 \pmod 4$ that can be written as the sum of two squares. It is easy to show that no integer $n \equiv 3 \pmod 4$ can be so represented. More generally we can ask: Which integers n can be represented as the sum of two squares and how many representations are there?

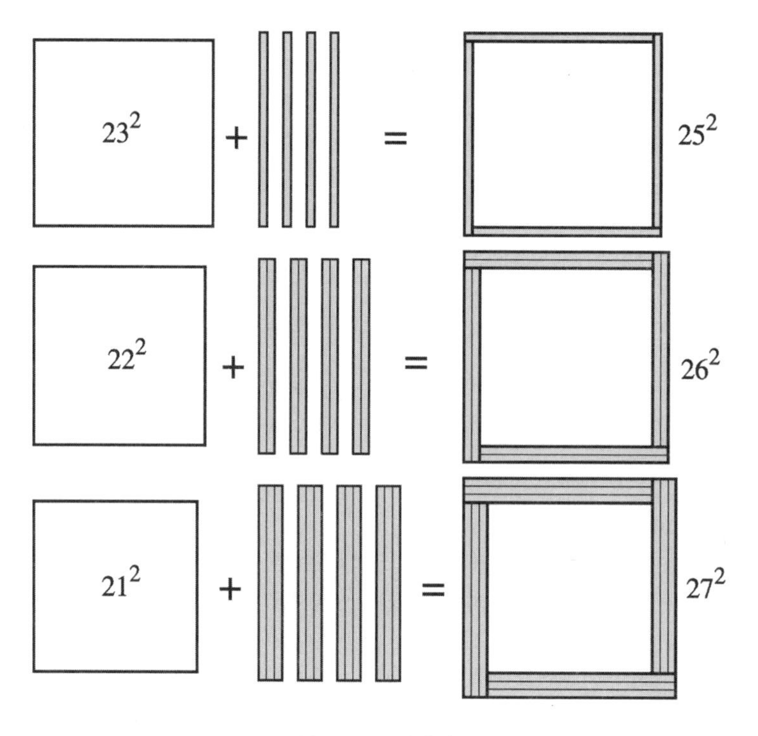

FIGURE 4.21

Let $r_2(n)$ denote the number of ways to represent an integer n as the sum of two squares (positive, negative, or zero), i.e., $r_2(n)$ denotes the number of solutions in integers (x, y) to the equation $x^2 + y^2 = n$. For example, $r_2(13) = 8$ since the solutions to $x^2+y^2 = 13$ are $(2,3)$, $(3,2)$, $(-2,3)$, $(3,-2)$, $(2,-3)$, $(-3,2)$, $(-2,-3)$, and $(-3,-2)$, as illustrated in Figure 3.1. Since $r_2(n) = 0$ whenever $n \equiv 3 \pmod 4$, $r_2(n)$ is a very erratic function, as illustrated in Figure 4.22 for $0 \le n \le 30$.

But we can ask: What is the average value of $r_2(k)$ for $1 \le k \le n$? We define $R_2(n)$ to be the number of solutions in integers to $x^2+y^2 \le n$, and then the average of $r_2(k)$ for $1 \le k \le n$ is

$$\frac{r_2(1) + r_2(2) + \cdots + r_2(n)}{n} = \frac{R_2(n)}{n}.$$

Computation of $N_2(n)$ and $N_2(n)/n$ yields Table 4.1.

TABLE 4.1

n	1	2	3	4	5	10	20	50	100
$R_2(n)$	5	9	9	13	21	37	69	161	317
$R_2(n)/n$	5	4.5	3	3.25	4.2	3.7	3.45	3.22	3.17

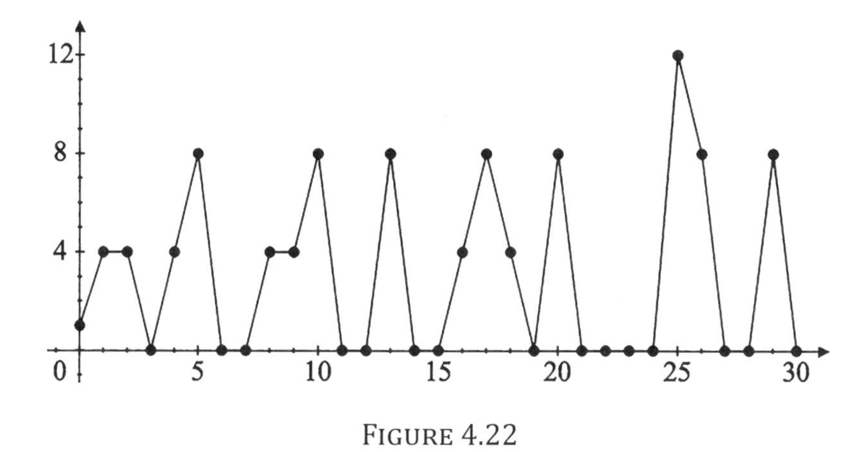

FIGURE 4.22

The average value of r_2 over all the positive integers is the limit of $R_2(n)/n$ as $n \to \infty$ if the limit exists. It does, and we have

Theorem 4.19. $\lim_{n\to\infty} \frac{R_2(n)}{n} = \pi$.

Proof. The proof is based on a geometric interpretation of $R_2(n)$ and is due to Gauss. $R_2(n)$ is the number of lattice points (points with integer coordinates) in or on the circle $x^2 + y^2 = n$. For example, $R_2(10) = 37$ since the circle centered at the origin with radius $\sqrt{10}$ contains 37 lattice points, as illustrated in Figure 4.23(a). If we draw a square with area 1 centered at each of the 37 lattice points, then the total area of the squares (in gray) is also $R_2(10)$. Thus we would expect the area of the

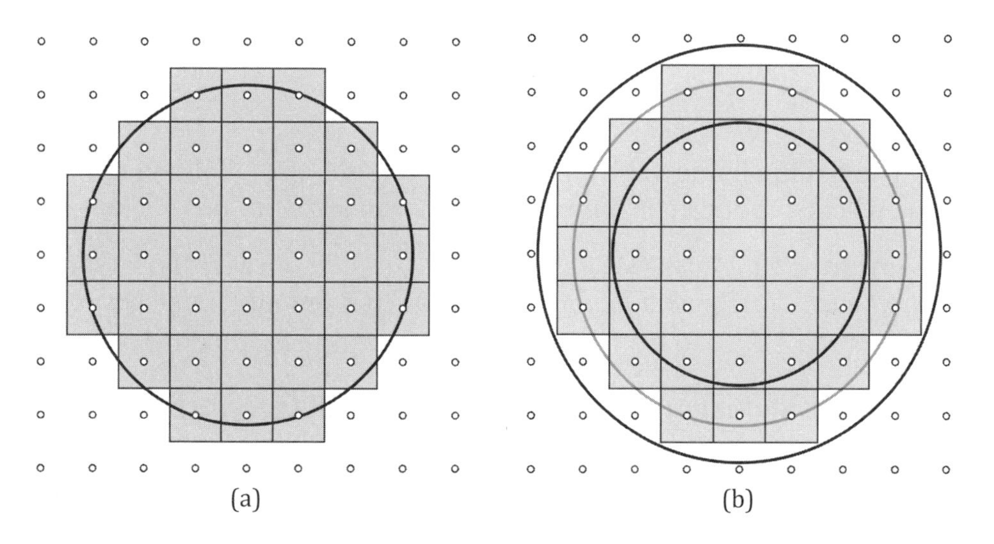

FIGURE 4.23

squares to be approximately the area of the circle, or in general, $R_2(n)$ to be approximately $\pi(\sqrt{n})^2 = \pi n$.

If we expand the circle with radius $\sqrt{n}$ by one-half the length of the diagonal ($\sqrt{2}/2$) of a square with area 1, then the expanded circle contains all the squares. If we contract the circle by the same amount, then the contracted circle is contained in the union of all the squares, as illustrated in Figure 4.23(b). Thus

$$\pi\left(\sqrt{n} - \sqrt{2}/2\right)^2 < R_2(n) < \pi\left(\sqrt{n} + \sqrt{2}/2\right)^2.$$

Dividing each term by n and applying the squeeze theorem for limits yields the desired result. □

We now return to the first question at the end of the introductory paragraph in this section: Which integers can be represented as the sum of two squares? In Table 4.2 we present the prime power factorizations for the positive integers considered in Figure 4.22 in two groups, those for which $r_2(k) > 0$ and those for which $r_2(k) = 0$.

TABLE 4.2

$r_2(k) > 0$		$r_2(k) = 0$	
$1 = 1$	$16 = 2^4$	$3 = 3$	$21 = 3 \cdot 7$
$2 = 2$	$17 = 17$	$6 = 2 \cdot 3$	$22 = 2 \cdot 11$
$4 = 2^2$	$18 = 2 \cdot 3^2$	$7 = 7$	$23 = 23$
$5 = 5$	$20 = 2^2 \cdot 5$	$11 = 11$	$24 = 2^3 \cdot 3$
$8 = 2^3$	$25 = 5^2$	$12 = 2^2 \cdot 3$	$27 = 3^3$
$9 = 3^2$	$26 = 2 \cdot 13$	$14 = 2 \cdot 7$	$28 = 2^2 \cdot 7$
$10 = 10$	$29 = 29$	$15 = 3 \cdot 5$	$30 = 2 \cdot 3 \cdot 5$
$13 = 13$		$19 = 19$	

The data in Table 4.2 leads to the statement of the following theorem, whose proof can be found in most number theory textbooks.

Theorem 4.20. *A positive integer is not the sum of two squares if and only if each of its prime factors congruent to* 3 (mod 4) *appears to an odd power.*

4.13. Pythagorean quadruples and Pythagorean boxes

A *Pythagorean quadruple* (PQ) is a quadruple (a, b, c, d) of positive integers such that $a^2 + b^2 + c^2 = d^2$. PQs can easily be constructed from PTs. For example, the PTs $(3, 4, 5)$ and $(5, 12, 13)$ yield the PQ $(3, 4, 12, 13)$. But this procedure only generates a few PQs. We present an extension of Euclid's formula that generates all the PQs [Sierpiński, 2003]. But first, we need the following lemma, from Problem 19 in Book III of *Arithmetica*, written by Diophantus of Alexandria. It was in the margin of one page in his copy of this book that Pierre de Fermat stated without proof his famous "Last Theorem." The *Arithmetica* was translated from Greek to Latin by Claude Gaspard Bachet de Méziriac in 1621. Below is the cover of an edition with commentary by Fermat published by Fermat's son Clément-Samuel de Fermat in 1670.

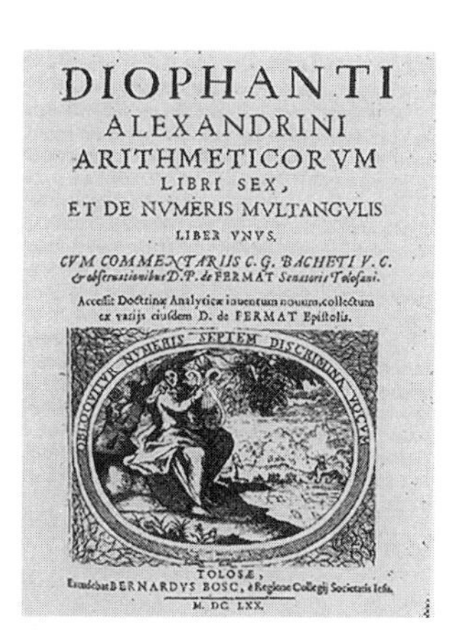

DIOPHANTI
ALEXANDRINI
ARITHMETICORVM
LIBRI SEX,
ET DE NVMERIS MVLTANGVLIS
LIBER VNVS.
CVM COMMENTARIIS C. G. BACHETI V. C.
& obseruationibus D. P. de FERMAT Senatoris Tolosani.
Accessit Doctrinæ Analyticæ inuentum nouum, collectum
ex varijs eiusdem D. de FERMAT Epistolis.

TOLOSÆ,
Excudebat BERNARDVS BOSC, è Regione Collegij Societatis Iesu.
M. DC. LXX.

Lemma 4.21. Diophantus of Alexandria's sum of squares identity. *If two positive integers are each sums of two squares, then their product is a sum of two squares in two different ways, i.e.,*

$$(a^2+b^2)(c^2+d^2) = (ac+bd)^2 + (ad-bc)^2 \tag{4.8}$$

and

$$(a^2+b^2)(c^2+d^2) = (ad+bc)^2 + (ac-bd)^2. \tag{4.9}$$

Proof. See Figure 4.24 for a proof of (4.8) for the case $ad > bc$ (the case $ad < bc$ is similar), using the Pythagorean theorem to show that $\left(\sqrt{a^2+b^2}\sqrt{c^2+d^2}\right)^2 = (ac+bd)^2 + (ad-bc)^2$. Exchanging c and d in the figure yields a proof of (4.9). □

We now state and prove an identity to generate PQs.

Theorem 4.22. *For integers m, n, p, and q,*

$$\begin{aligned}(m^2+n^2-p^2-q^2)^2 + (2mp+2nq)^2 + (2mq-2np)^2 \\ = (m^2+n^2+p^2+q^2)^2.\end{aligned}$$

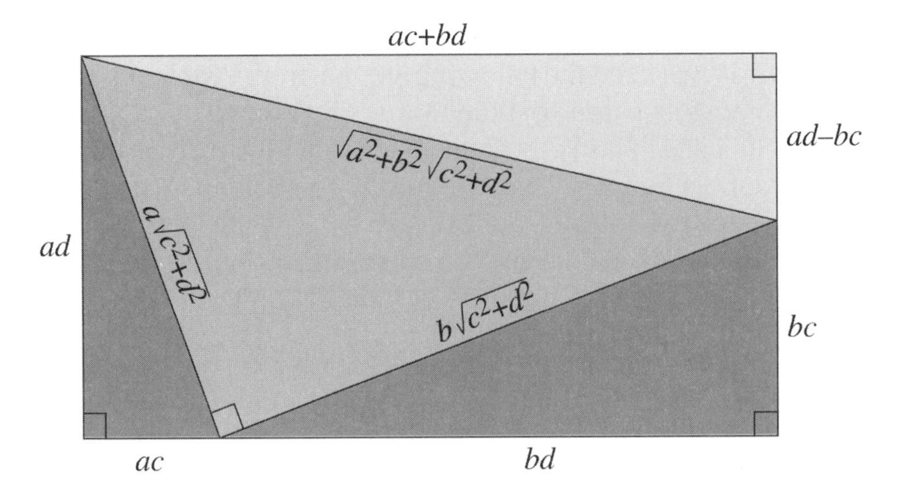

FIGURE 4.24

Proof. Setting $u = m^2 + n^2$ and $v = p^2 + q^2$ in Lemma 3.4 yields

$$(m^2 + n^2 - p^2 - q^2)^2 + 4\left(m^2 + n^2\right)\left(p^2 + q^2\right) = \left(m^2 + n^2 + p^2 + q^2\right)^2. \tag{4.10}$$

Applying (4.8) to $\left(m^2 + n^2\right)\left(p^2 + q^2\right)$ and multiplying by 4 yields

$$4\left(m^2 + n^2\right)\left(p^2 + q^2\right) = (2mp + 2nq)^2 + (2mq - 2np)^2,$$

which completes the proof. See [Spira, 1962] for a proof that this identity generates all PQs. □

A *Pythagorean box*, a three-dimensional analogue of a Pythagorean triangle, is a rectangular box whose edges and interior diagonal are positive integers, and hence it can be represented by a PQ (a, b, c, d). A *nearly cubic Pythagorean box* is one given by $(a, a, a + 1, d)$ or $(a, a + 1, a + 1, d)$. Examples include (1,2,2,3) and (6,6,7,11), as well as the degenerate box (0,0,1,1). We begin with a lemma relating one Pythagorean box to a larger one.

Lemma 4.23. $a^2 + b^2 + c^2 = d^2$ *if and only if*

$$(a + b + d)^2 + (b + c + d)^2 + (a + c + d)^2 = (a + b + c + 2d)^2.$$

Proof. In this proof we place equilateral triangular carpets in an equilateral triangular room. Place carpets with areas $(a + b + d)^2$, $(b + c + d)^2$, and $(a + c + d)^2$ in a room with area $(a + b + c + 2d)^2$, as shown in Figure 4.25 (here each small Δ and ∇ triangle has area 1, and we recall that each large triangle contains a square number of

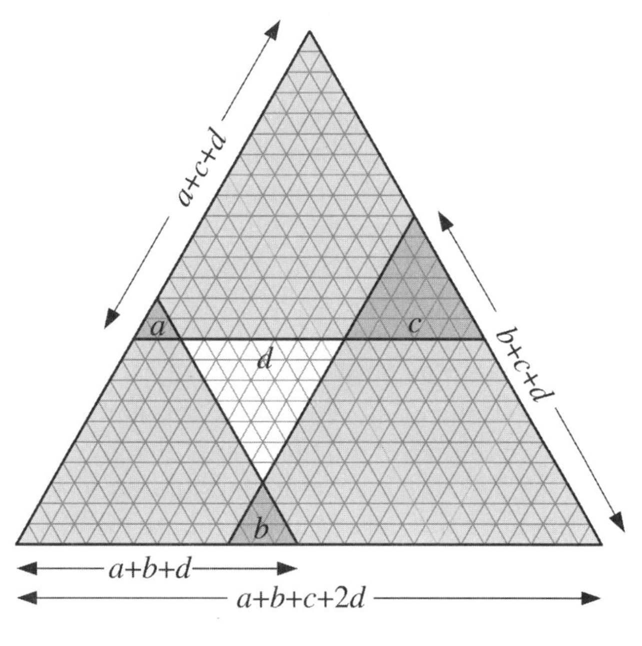

FIGURE 4.25

small $\triangle$ and $\triangledown$ triangles). The result now follows from the carpets theorem. $\square$

Theorem 4.24. *Infinitely many nearly cubic Pythagorean boxes exist.*

Proof. Repeated applications of Lemma 4.23 yield the following infinite sequence of PQs representing Pythagorean boxes:

$$(0,0,1,1) \to (1,2,2,3) \to (6,6,7,11) \to (23,24,24,41) \to (88,88,89,153) \to \cdots.$$

$\square$

4.14. Exercises

4.1 Show that if (a,b,c) is a PPT, then a and b have opposite parity and c is odd.

4.2 Show that the even leg of a PPT is always a multiple of 4, and that every multiple of 4 appears in some PPT.

4.3 Show that if (a,b,c) is a PPT, then $2c$ is a sum of two odd squares.

4.4 The PTs $(6,8,10)$ and $(27,36,45)$ share the interesting property that one leg is an integral cube while the other leg and the hypotenuse are consecutive triangular numbers. Are there other PTs with this property? If so, are any primitive?

4.5 In Exercise 2.8 we encountered the Fermat numbers, integers of the form $f_n = 2^{2^n} + 1$ for $n \geq 0$. Show that for $n \geq 1$, every Fermat number appears as the hypotenuse of some PPT.

4.6 Show that $T_m = k(k+1)$ is an oblong triangular number if and only if $(m, m+1, 2k+1)$ is an almost isosceles PPT.

4.7 Show that there are always two different PPTs with a given odd prime number as inradius. [Hint: Solve $p = n(m-n)$ for m and n.]

4.8 Show that there exist infinitely many PPTs whose area is the product of three consecutive integers.

4.9 In Chapter 1 we encountered the triangular numbers $\{T_k\} = \{1, 3, 6, 10, \ldots\}$ as sums of integers and the pyramidal numbers $\{\text{Pyr}_k\} = \{1, 5, 14, 30, \ldots\}$ as sums of squares of integers. These numbers appear in the perimeter P and area A of some primitive PΔ's, as Table 4.3 illustrates.

TABLE 4.3

PPT	P	A
(3,4,5)	$12 = 2 \cdot T_3$	$6 = 6 \cdot \text{Pyr}_1$
(5,12,13)	$30 = 2 \cdot T_5$	$30 = 6 \cdot \text{Pyr}_2$
(7,24,25)	$56 = 2 \cdot T_7$	$84 = 6 \cdot \text{Pyr}_3$

Are there other PΔ's with this property?

4.10 Prove the following identities for the inradius r, the three exradii r_a, r_b, r_c, the perimeter P, semiperimeter s, and the area k of a primitive PΔ:

(a) $r + r_a + r_b = r_c$,

(b) $r + r_a + r_b + r_c = P$,

(c) $\frac{1}{r} = \frac{1}{r_a} + \frac{1}{r_b} + \frac{1}{r_c}$,

(d) $r_a r_b = r r_c = K$ and hence $r r_a r_b r_c = K^2$,

(e) $r_a r_b + r_b r_c + r_c r_a = s^2$.

4.11 Prove that there are infinitely many Pythagorean boxes whose edges are integer squares. [Hint: Let (a, b, c) be a PT and show that the Pythagorean box with edges $(ab)^2$, $(bc)^2$, and $(ac)^2$ has interior diagonal $c^4 - a^2b^2$.]

4.12 Prove that if an integer is the sum of three positive integer squares, then so is its square.

4.13 *Three squares in arithmetic progression.* $(1, 25, 49)$ and $(289, 625, 961)$ are two examples of three squares in arithmetic progression. Are there others? [Hint: See Figure 4.26.]

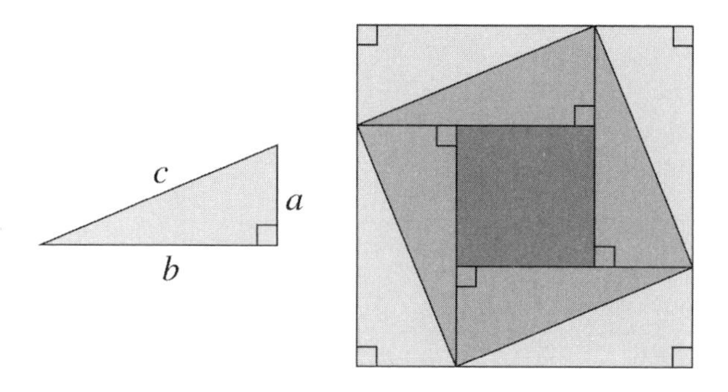

FIGURE 4.26

4.14 *The Diophantine equation* $x^2+y^2=z^3$. Sometimes the sum of two squares can be a cube, e.g., $5^3=10^2+5^2=2^2+11^2$ and $13^3=46^2+9^2=39^2+26^2$. Are there other solutions to $x^2+y^2=z^3$?

CHAPTER 5

Irrational Numbers

To attempt to apply rational arithmetic to a problem in geometry resulted in the first crisis in the history of mathematics. The two relatively simple problems—the determination of the diagonal of a square and that of the circumference of a circle—revealed the existence of new mathematical beings for which no place could be found within the rational domain.

Tobias Dantzig

While the primary objects of study in an elementary number theory course are the integers, real numbers formed from integers are often considered as well. The rational numbers are, as the name suggests, ratios of integers. Real numbers that are not rational—the irrational numbers—are perhaps of more interest than the rational numbers. This may follow from the definition of an irrational number, telling us what *it is not* (an irrational number is a real number that is not rational) rather than what *it is*.

We confine ourselves to the positive quadratic irrational numbers, those that are positive irrational roots of quadratic equations with integer coefficients. We begin with some visual demonstrations of the irrationality of simple quadratic irrationals such as $\sqrt{2}$, $\sqrt{3}$, $\sqrt{5}$, and the golden ratio. We illustrate rational approximations to these numbers using solutions to Pell equations and certain triangles with integer sides, and express some quadratic irrationals in terms of continued fractions.

5.1. The irrationality of $\sqrt{2}$

The discovery that $\sqrt{2}$ is irrational was probably due to the Pythagoreans, perhaps even Pythagoras himself. This may have occurred by observing that the hypotenuse of an isosceles right triangle is not commensurable with the legs (two non-zero real numbers are *commensurable* if and only if their ratio is rational). Since "irrational" means "not rational," virtually all proofs of irrationality are indirect, showing that the assumption that the number is rational leads to a contradiction.

Theorem 5.1. *$\sqrt{2}$ is irrational.*

Proof 1 ([Apostol, 2000]). If $\sqrt{2}$ were rational, then $\sqrt{2} = a/b$ in lowest terms for positive integers a and b. Then $2b^2 = a^2$, so that a right triangle with legs b and hypotenuse a is the smallest isosceles right triangle with integer sides. However (see Figure 5.1),

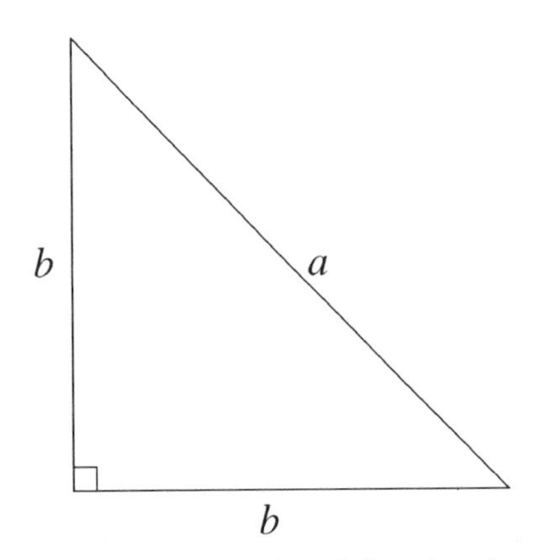

if this is an isosceles right triangle with integer sides,

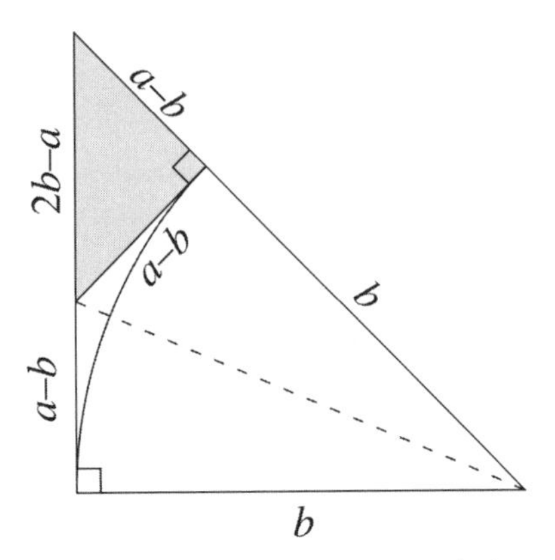

then there is a smaller one with the same property.

FIGURE 5.1

Indeed, if the sides of the larger triangle above are b, b, and a, then the sides of the smaller gray triangle on the right are positive integers $a - b$, $a - b$, and $2b - a$. Thus the assumption that $\sqrt{2} = a/b$ in lowest terms is false, and $\sqrt{2}$ is irrational. □

The above proof appears in [Bloom, 1995] as a one-sentence proof of the irrationality of $\sqrt{2}$: If $\sqrt{2}$ were rational, say $\sqrt{2} = a/b$ in lowest terms, then also $\sqrt{2} = (2b - a)/(a - b)$ in *lower* terms, giving a contradiction.

Here is a second proof, created by the American mathematician Stanley Tennenbaum (1927–2005) in the 1950s [Conway, 2005]. It employs the carpets theorem from Section 4.3.

Proof 2. Assume that $\sqrt{2}$ is rational, and write $\sqrt{2} = a/b$ in lowest terms for positive integers a and b. Then $a^2 = 2b^2 = b^2 + b^2$. So place two $b \times b$ carpets on the floor in an $a \times a$ room, as shown in Figure 5.2.

By the carpets theorem, the area $(2b - a)^2$ of the overlap (in dark gray) equals the combined area $2(a - b)^2$ of the uncovered floor (in white). But these squares have positive integer sides with $2b - a$ and $a - b$ smaller than a and b, respectively, a contradiction. Hence $\sqrt{2}$ is irrational. □

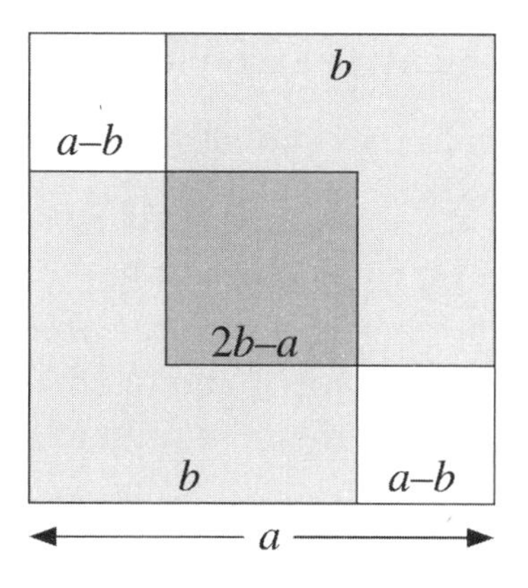

FIGURE 5.2

Hippasus and the irrationality of $\sqrt{2}$

Proving theorems in mathematics is not usually dangerous. Indeed, it is often quite rewarding, but in Pythagoras' time, the situation may have been different. Legend has it that Hippasus of Metapontum (c. 500 BCE) discovered the unexpected irrationality of $\sqrt{2}$, explained his result, and consequently was thrown overboard at sea by his fellow Pythagoreans. The discovery had destroyed the ideal of commensurability in geometry. Fortunately, history records no other such deaths as a result of proving theorems.

We conclude this section with four examples of rational and irrational numbers constructed from $\sqrt{2}$.

Example 5.1. *An irrational number to an irrational power may be rational.* To prove this statement, we need only give an example. If $\sqrt{2}^{\sqrt{2}}$ is rational, that is our example. If $\sqrt{2}^{\sqrt{2}}$ is irrational, then $\left(\sqrt{2}^{\sqrt{2}}\right)^{\sqrt{2}} = 2$ is our example. □

Example 5.2. *An irrational number to an irrational power may be irrational.* If $\sqrt{2}^{\sqrt{2}}$ is irrational, that is our example. If $\sqrt{2}^{\sqrt{2}}$ is rational, then $\sqrt{2}^{\sqrt{2}+1} = \sqrt{2}^{\sqrt{2}} \cdot \sqrt{2}$ is our example (since the sum or product of a rational number and an irrational number is irrational). □

An irrational number to a rational power may be rational ($\sqrt{2}^{2}$) or irrational ($\sqrt{2}^{1}$). The case of a rational number to an irrational power is examined in Exercise 5.2.

We proved the statements in Examples 5.1 and 5.2 without knowing whether $\sqrt{2}^{\sqrt{2}}$ is rational or irrational. We used only the fact that $\sqrt{2}$ is irrational. In fact, $\sqrt{2}^{\sqrt{2}}$ is irrational, as it is the square root of the *Gelfond-Schneider constant* $2^{\sqrt{2}}$, which is known to be transcendental. A (possibly complex) number is *transcendental* if it is not algebraic; a number is algebraic if it is the root of a non-zero polynomial with integer coefficients; all (real) transcendental numbers are irrational. However, it is unknown whether $\sqrt{2}^{\sqrt{2}^{\sqrt{2}}}$ is rational or irrational.

Hilbert and the Gelfond–Schneider constant $2^{\sqrt{2}}$

Early in Chapter 3 we mentioned Hilbert and the address with 23 problems that he gave at the Second International Congress of Mathematicians in Paris in 1900. In his seventh problem, he conjectured that "the expression a^b, for algebraic base a (not equal to 0 or 1) and an irrational algebraic exponent b, e.g., $2^{\sqrt{2}}$, always represents a transcendental or at least an irrational number." The conjecture was proved by A. O. Gelfond and T. Schneider independently in 1934. However, it is still unknown whether a^b is transcendental when both a and b are transcendental.

Example 5.3. *A tower of square roots of* 2. What is the meaning of an expression such as

$$\sqrt{2}^{\sqrt{2}^{\sqrt{2}^{\sqrt{2}^{\cdot^{\cdot^{\cdot}}}}}} ?$$

The ellipsis "$\cdot^{\cdot^{\cdot}}$" indicates infinitely many copies of $\sqrt{2}$ as exponents, so we consider the limit of a sequence. Set $x_1 = \sqrt{2}$, define $x_{n+1} = \sqrt{2}^{x_n}$ for $n \geq 1$, and then the question becomes: does the sequence $\{x_n\}$ converge, and, if so, to what? The sequence converges since it is increasing (the function $f(x) = \sqrt{2}^{x}$ is increasing) and bounded above by 2 (since $\sqrt{2} < 2$ we have

$$x_n = \sqrt{2}^{\sqrt{2}^{\cdot^{\cdot^{\cdot^{\sqrt{2}^{\sqrt{2}^{\sqrt{2}}}}}}}} < \sqrt{2}^{\sqrt{2}^{\cdot^{\cdot^{\cdot^{\sqrt{2}^{\sqrt{2}^{2}}}}}}} = \sqrt{2}^{\sqrt{2}^{\cdot^{\cdot^{\cdot^{\sqrt{2}^{2}}}}}} = \cdots = \sqrt{2}^{2} = 2).$$

Hence the sequence $\{x_n\}$ has a limit $L \leq 2$, and

$$L = \lim_{n\to\infty} x_{n+1} = \lim_{n\to\infty} \sqrt{2}^{x_n} = \sqrt{2}^{\lim_{n\to\infty} x_n} = \sqrt{2}^{L}.$$

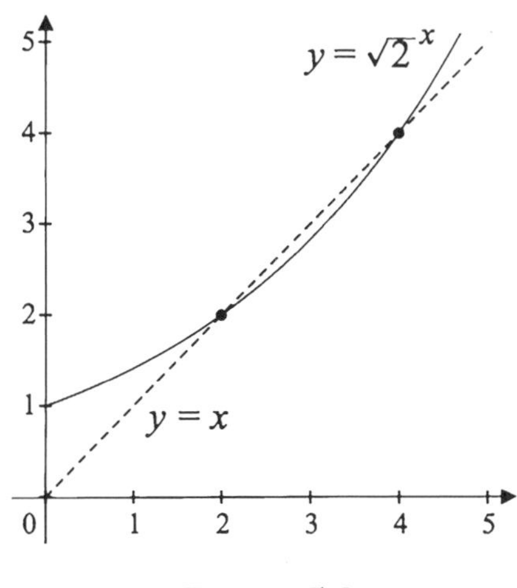

FIGURE 5.3

By inspection (see Figure 5.3) the equation $L = \sqrt{2}^L$ has solutions $L = 2$ and $L = 4$. Thus the limit is 2 since we know that $L \leq 2$. □

Example 5.4. *François Viète's remarkable formula for π.* In 1583 the French mathematician François Viète (1540–1603) discovered the following infinite product for π employing only 2 and $\sqrt{2}$:

$$\frac{2}{\pi} = \frac{\sqrt{2}}{2} \cdot \frac{\sqrt{2+\sqrt{2}}}{2} \cdot \frac{\sqrt{2+\sqrt{2+\sqrt{2}}}}{2} \cdot \frac{\sqrt{2+\sqrt{2+\sqrt{2+\sqrt{2}}}}}{2} \cdots .$$

We prove this formula as Viète did, with a geometric argument based on ratios of the areas of regular polygons inscribed in the unit circle. First notice that each term in the infinite product is a cosine obtained from $\cos \pi/4 = \sqrt{2}/2$ by repeated use of the half-angle formula $\cos(\theta/2) = \sqrt{(1+\cos\theta)/2} = \frac{1}{2}\sqrt{2+2\cos\theta}$. It follows that the terms in the product are given by $\cos(\pi/2^n)$ for $n \geq 2$.

Let $V_n = \cos(\pi/4) \cdot \cos(\pi/8) \cdots \cos(\pi/2^n)$ for $n \geq 2$. To show that the limit of V_n as n tends to infinity is $2/\pi$, we inscribe regular polygons with 2^n sides in the unit circle. It is easy to show that the area of a regular polygon with k sides inscribed in the unit circle has area $(k/2)\sin(2\pi/k)$. So if A_n denotes the area of a polygon with 2^n sides, then $A_n = 2^{n-1}\sin(\pi/2^{n-1})$ for $n \geq 2$. Hence the ratio of the areas of two successive such polygons is

$$\frac{A_n}{A_{n+1}} = \frac{2^{n-1}\sin(\pi/2^{n-1})}{2^n \sin(\pi/2^n)}$$
$$= \frac{2^n \sin(\pi/2^n)\cos(\pi/2^n)}{2^n \sin(\pi/2^n)}$$
$$= \cos(\pi/2^n)$$

for $n \geq 2$. Thus

$$V_n = \frac{A_2}{A_3} \cdot \frac{A_3}{A_4} \cdots \frac{A_{n-1}}{A_n} \cdot \frac{A_n}{A_{n+1}} = \frac{A_2}{A_{n+1}}$$

so that $\lim_{n\to\infty} V_n = 2/\pi$ since $A_2 = 2$ (the area of a square inscribed in the unit circle) and $\lim_{n\to\infty} A_{n+1} = \pi$ (the limiting area of the inscribed polygons is the area of the unit circle). □

See Exercise 5.7 for a related result involving $\sqrt{2}$.

5.2. Rational approximations to $\sqrt{2}$: Pell equations

Tennenbaum's proof that $\sqrt{2}$ is irrational (proof 2 of Theorem 5.1) can be modified to provide a procedure for generating rational approximations to $\sqrt{2}$. The procedure is very similar to the approach to solving Pell's equations in Chapter 3. We set $x = 2b - a$ and $y = a - b$ in Figure 5.2 in the preceding section and employ the carpets theorem to prove the next theorem.

Theorem 5.2. *Let x and y be positive integers. Then*

(5.1) $\quad (x + 2y)^2 - 2(x + y)^2 = \pm 1$ *if and only if* $x^2 - 2y^2 = \mp 1$.

Proof. See Figure 5.4 and compute the area of the large square in two ways to yield $(x + 2y)^2 = 2(x + y)^2 - x^2 + 2y^2$. Hence

$$x^2 - 2y^2 = 2(x + y)^2 - (x + 2y)^2,$$

from which the desired result follows. □

Set

(5.2) $\quad x_{n+1} = x_n + 2y_n \quad \text{and} \quad y_{n+1} = x_n + y_n$

in (5.1) yielding $x_{n+1}^2 - 2y_{n+1}^2 = \pm 1$ if and only if $x_n^2 - 2y_n^2 = \mp 1$. Now set $(x_1, y_1) = (1, 1)$, and hence $x_n^2 - 2y_n^2 = (-1)^n$ for all $n \geq 1$. Division by y_n^2 yields for all $n \geq 1$,

(5.3) $$\left(\frac{x_n}{y_n}\right)^2 - 2 = \frac{(-1)^n}{y_n^2}.$$

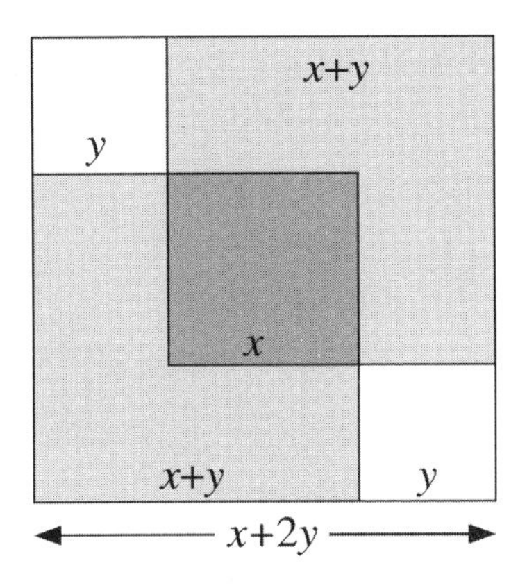

FIGURE 5.4

Thus x_n/y_n will approximate $\sqrt{2}$ (larger than $\sqrt{2}$ when n is even, smaller when n is odd) with an absolute error less than $1/2y_n^2$. This error bound follows from

$$\frac{1}{y_n^2} = \left(\frac{x_n}{y_n} + \sqrt{2}\right)\left|\frac{x_n}{y_n} - \sqrt{2}\right| > 2\left|\frac{x_n}{y_n} - \sqrt{2}\right|$$

since $\frac{x_n}{y_n} + \sqrt{2} \geq 1 + \sqrt{2} > 2$. See Table 5.1 for the numerators and denominators of the first few approximations to $\sqrt{2}$ generated by Theorem 5.2 with $(x_1, y_1) = (1, 1)$.

TABLE 5.1

n	1	2	3	4	5	6	7
x_n	1	3	7	17	41	99	239
y_n	1	2	5	12	29	70	169

Example 5.5. *A geometric interpretation of the error.* A triangle (y_n, y_n, x_n) with side lengths from Table 5.1 is isosceles and almost a right triangle since $x_n^2 = y_n^2 + y_n^2 + (-1)^n$. If α_n denotes the size of the angle opposite the longest side x_n, then the law of cosines yields $x_n^2 = 2y_n^2 - 2y_n^2 \cos \alpha_n$ so that $\cos \alpha_n = -(-1)^n/(2y_n^2)$. Thus the error term $(-1)^n/y_n^2$ in (5.3) is $-2 \cos \alpha_n$. □

5.3. Rational approximations to $\sqrt{2}$: Almost isosceles PPTs

In Theorem 4.8 we encountered a relationship between almost isosceles PPTs and the Pell equations $x^2 - 2y^2 = \pm 1$. Geometrically, the right triangle whose sides are the elements of an almost isosceles PPT is "almost similar" to an isosceles right triangle. Consequently, we can use ratios of functions of the sides of such PPTs to approximate $\sqrt{2}$, via the following theorem.

Theorem 5.3. *Let* (a, b, c) *with* $|a - b| = 1$ *be an almost isosceles PPT. Then* $(x, y) = (a + b, c)$ *is a solution to* $x^2 - 2y^2 = -1$ *and* $(x, y) = (a + b + 2c, a + b + c)$ *is a solution to* $x^2 - 2y^2 = +1$*. Hence* $(a + b)/c$ *and* $(a + b + 2c)/(a + b + c)$ *are rational approximations to* $\sqrt{2}$*.*

Proof. We consider the case $b = a + 1$; the case $a = b + 1$ is similar. If (a, b, c) is an almost isosceles PΔ, then computing the area of the square in Figure 5.5 in two ways yields $c^2 = 4 \cdot \frac{1}{2} a(a+1) + 1 = 2a\,(a + 1) + 1$.

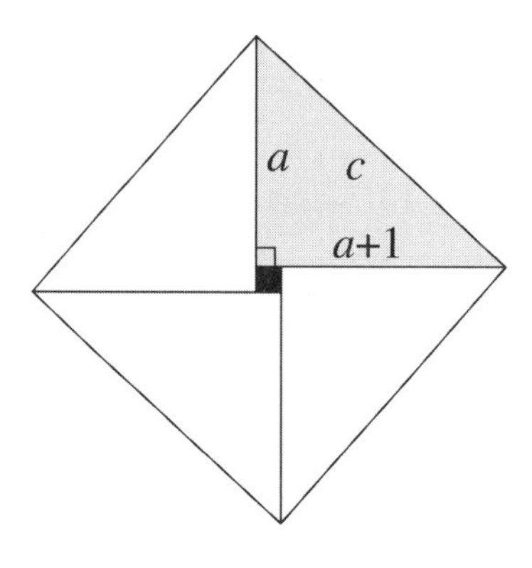

FIGURE 5.5

Completing the square produces $(2a+1)^2 - 2c^2 = -1$, or $(a+b)^2 - 2c^2 = -1$. Thus $(x, y) = (a + b, c)$ is a solution to $x^2 - 2y^2 = -1$, and consequently $(a + b)/c$ is an approximation to $\sqrt{2}$. From (5.2) we conclude that $(x, y) = (a + b + 2c, a + b + c)$ is a solution to $x^2 - 2y^2 = +1$, and consequently $(a + b + 2c)/(a + b + c)$ is also an approximation to $\sqrt{2}$. □

In Section 4.5 we encountered the first four almost isosceles PPTs: $(3, 4, 5)$, $(21, 20, 29)$, $(119, 120, 169)$, and $(697, 696, 985)$. These yield the following approximations to $\sqrt{2}$: From $(a+b)/c$ we have $7/5$, $41/29$, $239/169$, and $1393/985$ (see Table 3.4); and from $(a + b + 2c)/(a + b + c)$ we have $17/12$, $99/70$, $577/408$, and $3363/2378$ (see Table 3.3). The numerators and denominators of these approximations also appear in Table 5.1.

5.4. The irrationality of $\sqrt{3}$ and $\sqrt{5}$

We mimic the two proofs of the irrationality of $\sqrt{2}$ to give two proofs that $\sqrt{3}$ is irrational.

Theorem 5.4. *$\sqrt{3}$ is irrational.*

Proof 1. Assume that $\sqrt{3}$ is rational, and write $\sqrt{3} = a/b$ in lowest terms for positive integers a and b. Then $a^2 = 3b^2$ or, equivalently, $a^2 + b^2 = (2b)^2$. Now draw a right triangle with hypotenuse $2b$ and legs a and b, as shown in Figure 5.6.

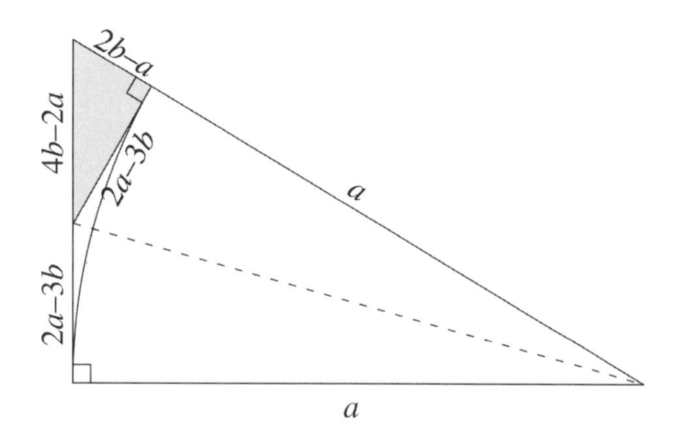

FIGURE 5.6

We now construct the smaller right triangle shaded gray with integer hypotenuse $4b - 2a$ and integer legs $2b - a$ and $2a - 3b$, so that $(2a - 3b)^2 = (4b - 2a)^2 - (2b - a)^2 = 3(2b - a)^2$. Thus $\sqrt{3} = (2a - 3b)/(2b - a)$, a contradiction since $2a-3b$ and $2b-a$ are smaller than a and b, respectively. Hence $\sqrt{3}$ is irrational. □

Proof 2. Assume that $\sqrt{3}$ is rational, and write $\sqrt{3} = a/b$ in lowest terms for positive integers a and b. Then $a^2 = 3b^2 = b^2 + b^2 + b^2$. We place three equilateral triangular carpets with side b in an equilateral triangular room with side a, as shown in Figure 5.7.

The carpets overlap in three equilateral triangles with side $2b - a$ (shaded dark gray) and leave uncovered an equilateral triangle portion of the floor with side $2a - 3b$ (in white). By the carpets theorem, the combined area $3k(2b - a)^2$ equals the area $k(2a - 3b)^2$ of the floor ($k = \sqrt{3}/4$ whether $\sqrt{3}$ is rational or not). Thus $\sqrt{3} = (2a - 3b)/(2b - a)$, a contradiction since $2a - 3b$ and $2b - a$ are smaller than a and b, respectively. Hence $\sqrt{3}$ is irrational. □

Example 5.6. *Rational approximations to $\sqrt{3}$.* Analogous to what we did in Sections 5.2 and 5.3 to approximate $\sqrt{2}$, we can use solutions to the Pell equation $x^2 - 3y^2 = 1$ to approximate $\sqrt{3}$. If (x_n, y_n) is a solution

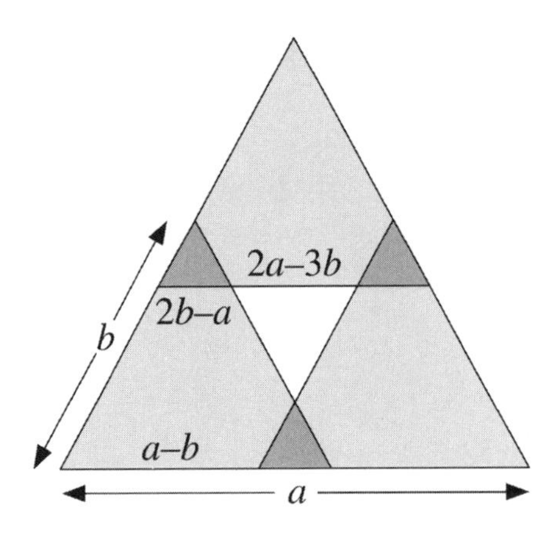

FIGURE 5.7

to $x^2 - 3y^2 = 1$ (see Section 3.5), then when $n \geq 1$ we have

$$\left(\frac{x_n}{y_n}\right)^2 - 3 = \frac{1}{y_n^2}.$$

Thus x_n/y_n yields an approximation to $\sqrt{3}$ (larger than $\sqrt{3}$) with an absolute error less than $1/3y_n^2$. The error estimate follows from

$$\frac{1}{y_n^2} = \left(\frac{x_n}{y_n} + \sqrt{3}\right)\left(\frac{x_n}{y_n} - \sqrt{3}\right) > 3\left(\frac{x_n}{y_n} - \sqrt{3}\right)$$

since $\frac{x_n}{y_n} + \sqrt{3} \geq \sqrt{3} + \sqrt{3} > 3$. In this case we do not obtain approximations less than $\sqrt{3}$, since the Pell equation $x^2 - 3y^2 = -1$ has no integer solutions (to see why, examine both sides of the equation modulo 3 and recall Example 3.2).

Analogous to our use of almost isosceles PPTs to approximate $\sqrt{2}$ in the previous section, we can use almost equilateral Heronian triangles from Example 3.10 to approximate $\sqrt{3}$. In that example we showed that if (x_n, y_n) is a solution to $x^2 - 3y^2 = 1$, then $(2x_n - 1, 2x_n, 2x_n + 1)$ is an almost equilateral Heronian triangle with semiperimeter $s = 3x_n$ and altitude (to the even side $2x_n$) $h = 3y_n$. Thus the ratio of semiperimeter to altitude is x_n/y_n and approximates $\sqrt{3}$. □

We conclude this section with a single proof that $\sqrt{5}$ is irrational.

Theorem 5.5. *$\sqrt{5}$ is irrational.*

Proof. Assume that $\sqrt{5}$ is rational, and write $\sqrt{5} = a/b$ in lowest terms for positive integers a and b. Then $a^2 = 5b^2$, or equivalently, $a^2 = b^2 + (2b)^2$. Now draw a right triangle with hypotenuse a and legs b and $2b$, as shown in Figure 5.8.

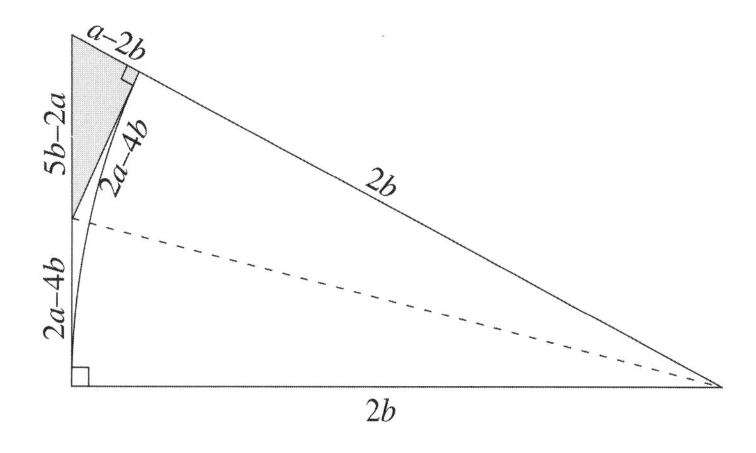

FIGURE 5.8

We now construct the smaller right triangle shaded gray with integer hypotenuse $5b - 2a$ and integer legs $a - 2b$ and $2a - 4b$, so that $(5b-2a)^2 = (a-2b)^2 + (2a-4b)^2 = 5(2b-a)^2$. Thus $\sqrt{5} = (5b-2a)/(a-2b)$, a contradiction since $5b-2a$ and $a-2b$ are smaller than a and b, respectively. Hence $\sqrt{5}$ is irrational. □

For another proof of the irrationality of $\sqrt{5}$ utilizing five pentagonal carpets in a pentagonal room, see [Miller and Montague, 2012]. See Example 6.12 for rational approximations to $\sqrt{5}$ using Fibonacci and Lucas numbers.

5.5. The irrationality of $\sqrt{d}$ for non-square d

We now modify the first proof of the irrationality of $\sqrt{2}$ to show that $\sqrt{d}$ is irrational when d is not the square of an integer. In the proof we interpret $\sqrt{d}$ as the slope of a line through the origin, as shown in Figure 5.9.

Theorem 5.6. *If d is not the square of an integer, then $\sqrt{d}$ is irrational.*

Proof. Assume $\sqrt{d} = a/b$ in lowest terms. Then the point on the line $y = \sqrt{d}x = (a/b)x$ closest to the origin with integer coordinates is (b, a). Let k be the greatest integer less than $\sqrt{d}$, so that $k < \sqrt{d} < k+1$. Then the point with integer coordinates $(a-kb, db-ka)$ lies on the line and is closer to the origin since $(a/b)(a-kb) = a^2/b - ka = db - ka$, and $k < a/b < k+1$ implies $0 < a-kb < b$ and $0 < db-ka < a$. Thus we have a contradiction, and $\sqrt{d}$ is irrational. When d is the square of an integer, i.e., $d = k^2$, then $a/b = k/1$ and $a - kb = 0 = db - ka$. □

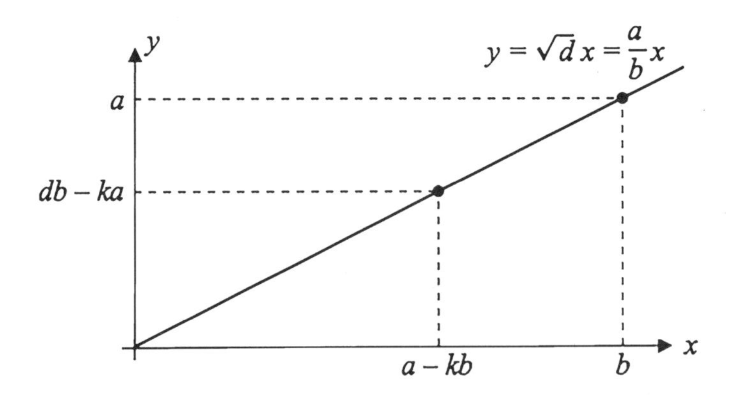

FIGURE 5.9

5.6. The golden ratio and the golden rectangle

The *golden ratio* (also known as the *golden section*, the *golden mean*, and the *divine proportion*) is a ratio that occurs in a variety of mathematical settings. Euclid provides a definition and a construction in Book VI of the *Elements*. Paraphrasing Euclid, we say that a line segment AB is divided by a point C into the golden ratio if $AB/AC = AC/BC$, or $\frac{a+b}{a} = \frac{a}{b}$. See Figure 5.10(a).

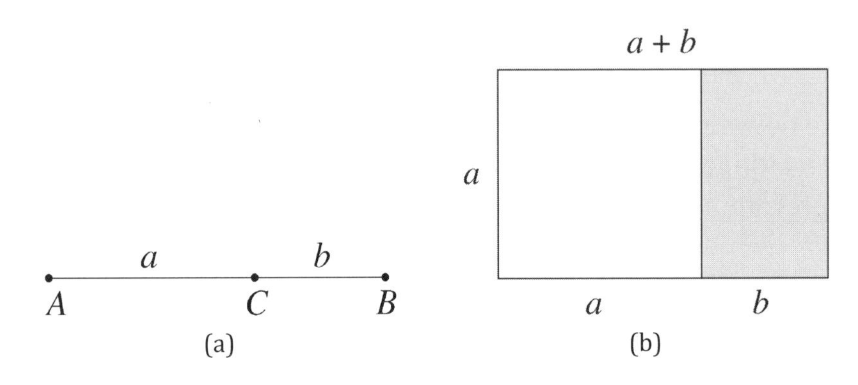

FIGURE 5.10

Using the segment AB as the base of a rectangle with height equal to AC yields a *golden rectangle* (see Figure 5.10(b)), a rectangle that has the property that when a square is cut off, the remaining rectangle (in gray) is similar to the original.

Using φ to denote the golden ratio, we have

$$\varphi = \frac{a}{b} = \frac{a+b}{a} = 1 + \frac{b}{a} = 1 + \frac{1}{\varphi},$$

so that $\varphi^2 = \varphi + 1$, and thus φ is the positive root of $x^2 - x - 1 = 0$.

To show that φ is irrational, we need only solve $x^2 - x - 1 = 0$ for its positive root using the quadratic formula: $\varphi = (1 + \sqrt{5})/2$. But it is not necessary to solve the quadratic equation, as the following theorem shows.

Theorem 5.7. *φ is irrational.*

Proof. Assume φ is rational, and write $\varphi = a/b$ in lowest terms for positive integers a and b. Then $\varphi = \frac{a}{b} = \frac{a+b}{a}$ implies that $a(a-b) = a^2 - ab = b^2$. Thus $\varphi = \frac{a}{b} = \frac{b}{a-b}$, a contradiction, since b and $a-b$ are positive integers smaller than a and b, respectively. Hence φ is irrational. See Figure 5.11. □

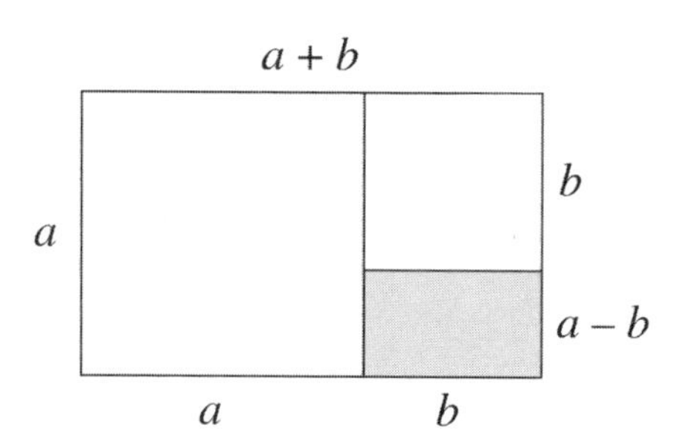

FIGURE 5.11

In addition, using the quadratic formula to obtain $\varphi = (1 + \sqrt{5})/2$ from $\varphi = 1 + \frac{1}{\varphi}$ is not necessary either. In the following theorem we dissect a square with area 5 to obtain the result.

Theorem 5.8. *$\varphi = 1 + \frac{1}{\varphi}$ with $\varphi > 0$ implies $\varphi = \frac{1+\sqrt{5}}{2}$.*

Proof. See Figure 5.12 to see that $(2\varphi - 1)^2 = 5$, from which the result follows. □

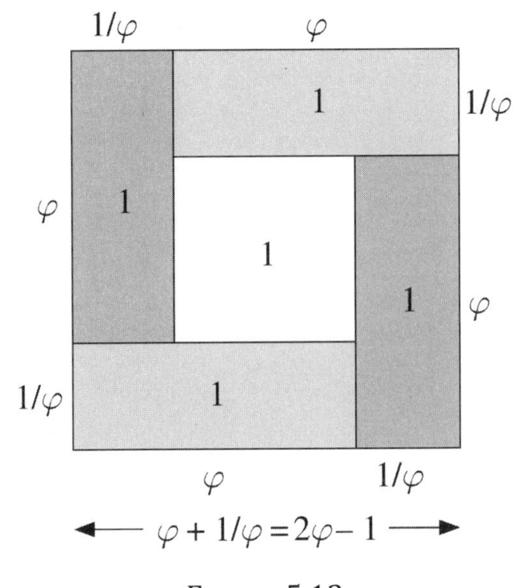

FIGURE 5.12

5.7. The golden ratio and the regular pentagon

The golden ratio φ may have been the second number shown to be irrational by the Pythagoreans—not from a golden rectangle, but from a regular pentagon. Since the diagonal of a square with side 1 is the irrational number $\sqrt{2}$, it is natural to also consider the length of the diagonal of a regular pentagon with side 1. We now show that the length of this diagonal is φ. See Figure 5.13.

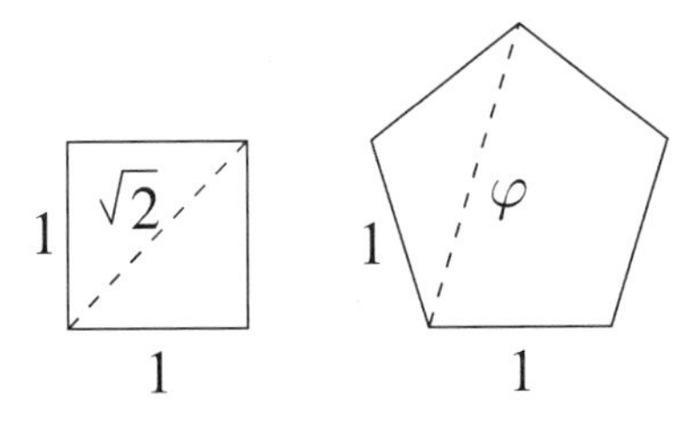

FIGURE 5.13

Let x denote the length of the diagonal of a regular pentagon with side 1, as shown in Figure 5.14(a). Two diagonals from a common vertex and the opposite side form an isosceles triangle (sometimes called a *golden triangle*), which can be partitioned into two smaller triangles, as illustrated in Figure 5.14(b).

The angles labeled α in Figure 5.14(b) are equal since each subtends an arc equal to 1/5 the circumference of the circumscribing circle. Thus the striped triangle is also isosceles, $\beta = 2\alpha$, and so $\alpha = 36^\circ$ since the

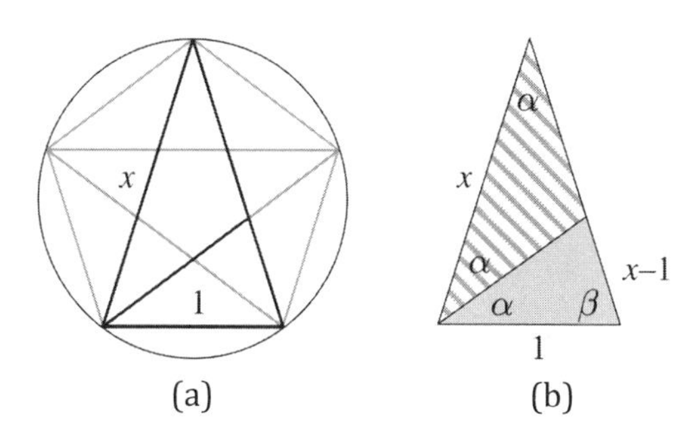

FIGURE 5.14

sum of the angles in the original triangle yields $5\alpha = 180°$. It now follows that the unlabeled angle in the gray triangle is $\beta = 72°$, and hence the gray triangle is also isosceles and similar to the original triangle. Thus $x/1 = 1/(x-1)$, and so x is the positive root of $x^2 - x - 1 = 0$. Thus $x = \varphi$ as claimed.

The regular pentagon can also be used to give another proof of Theorem 5.7 (φ is irrational). Assume that φ is rational and write $\varphi = a/b$ in lowest terms for positive integers a and b. Draw a regular pentagon with sides b and diagonals a as shown in Figure 5.15(a).

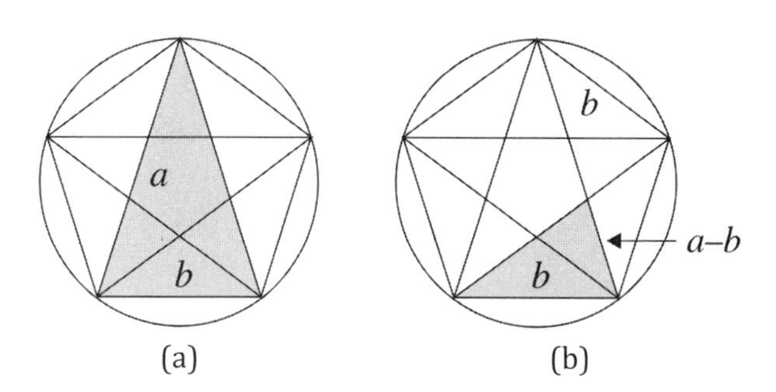

FIGURE 5.15

As shown in Figure 5.14(b), the gray triangle in Figure 5.15(b) is similar to the gray triangle in Figure 5.15(a), and hence $\varphi = b/(a-b)$, a contradiction, since b and $a - b$ are positive integers smaller than a and b, respectively. Hence φ is irrational.

5.8. Periodic continued fractions

An *infinite simple continued fraction* is an expression of the form

$$[a_0, a_1, a_2, a_3, ...] = a_0 + \cfrac{1}{a_1 + \cfrac{1}{a_2 + \cfrac{1}{a_3 + \ddots}}},$$

where a_0 is an integer and $a_1, a_2, a_3, ...$ are positive integers. When a block of a_k's repeats over and over, the continued fraction is *periodic*. For example,

$$[\,\overline{a}\,] = [a, a, a, ...] = a + \cfrac{1}{a + \cfrac{1}{a + \cfrac{1}{a + \ddots}}}$$

and

$$[\overline{a, b}] = [a, b, a, b, ...] = a + \cfrac{1}{b + \cfrac{1}{a + \cfrac{1}{b + \ddots}}}$$

are periodic (the vinculum or overbar indicates the repeating block of integers) with periods 1 and 2, respectively.

In most number theory texts you can find theorems that state that every infinite simple continued fraction represents an irrational number, and those that are periodic represent positive quadratic irrationals—numbers of the form $p + q\sqrt{k}$, where p and $q \neq 0$ are rational and k is a non-square integer (i.e., $p + q\sqrt{k}$ is a positive irrational root of a quadratic equation with integer coefficients).

We now evaluate the irrational number represented by the periodic continued fraction $[\,\overline{a, b}\,]$ (note that $[\,\overline{a}\,]$ with period 1 is a special case of $[\,\overline{a, b}\,]$ with period 2).

Theorem 5.9. *For positive integers a and b,*

$$[\overline{a, b}] = \frac{1}{2}[a + \sqrt{a^2 + 4(a/b)}].$$

Proof. Let $x = [\overline{a, b}] = a + \frac{1}{b+\frac{1}{x}}$. Then $bx + 1 = ab + \frac{a}{x} + 1$ so that $x = a + \frac{a}{bx}$, and hence $x + \frac{a}{bx} = 2x - a$. We now evaluate the area of a square with side $2x - a$ in two different ways (see Figure 5.16). Thus $(2x - a)^2 = a^2 + 4(a/b)$ so that $x = \frac{1}{2}[a + \sqrt{a^2 + 4(a/b)}]$. □

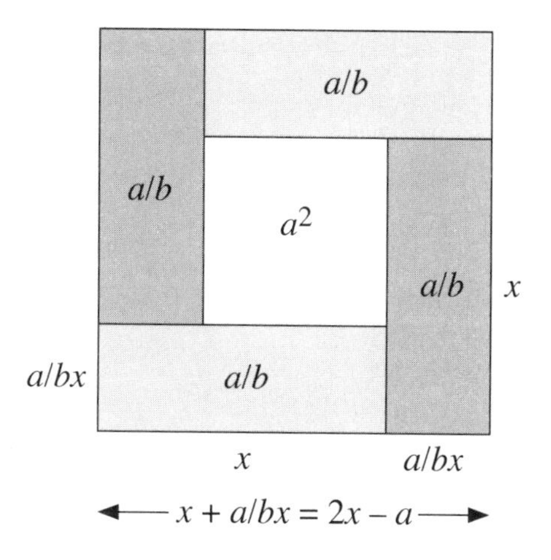

FIGURE 5.16

Example 5.7. *The golden, silver, and bronze ratios.* If we set $a = b = 1$ in Theorem 5.9 we have $[\overline{1}] = \frac{1}{2}(1 + \sqrt{5}) = \varphi$, that is, the golden ratio has what may well be the simplest periodic continued fraction:

$$\varphi = [\,\overline{1}\,] = 1 + \cfrac{1}{1 + \cfrac{1}{1+\ddots}}.$$

In general, the simple periodic continued fractions $[\overline{n}]$ with period 1 are given by

$$\varphi_n = [\,\overline{n}\,] = n + \cfrac{1}{n + \cfrac{1}{n+\ddots}} = \frac{1}{2}(n + \sqrt{n^2 + 4})$$

(where $\varphi_1 = \varphi$). These are the so-called *metallic ratios*, and correspond to ratios of the dimensions of rectangles analogous to the golden rectangle. For example, a *silver rectangle* is one with the property that when two squares are cut off (see Figure 5.17(a)), the remaining rectangle (in gray) is similar to the original.

Thus the *silver ratio* equals $\frac{x}{1} = \frac{2x+1}{x} = 2 + \frac{1}{x} =[\overline{2}] = \varphi_2 = 1 + \sqrt{2}$. Similarly, a *bronze rectangle* is one with the property that when three squares are cut off (see Figure 5.17(b)), the remaining rectangle (in gray) is similar to the original. Thus the *bronze ratio* equals $\frac{x}{1} = \frac{3x+1}{x} = 3 + \frac{1}{x} =[\overline{3}] = \varphi_3 = \frac{1}{2}(3 + \sqrt{13})$. The remaining metallic ratios φ_n are obtained similarly. □

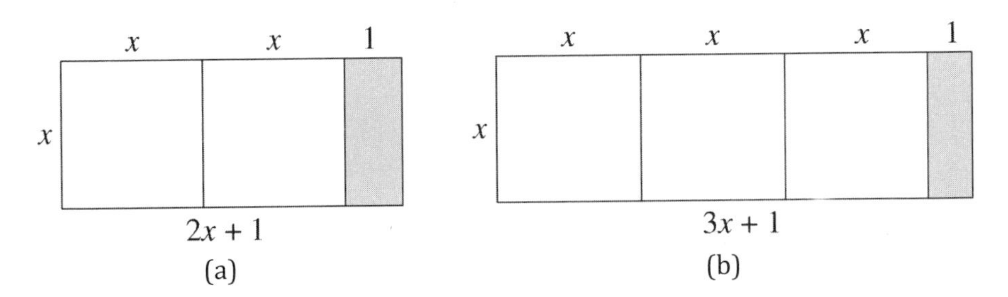

FIGURE 5.17

Theorem 5.9 also yields periodic continued fractions for some square roots. For example, for n a positive integer,

$$[\overline{2n}] = n + \sqrt{n^2+1} = 2n + \cfrac{1}{2n + \cfrac{1}{2n+\ddots}}$$

and subtracting n from both sides yields

$$\sqrt{n^2+1} = n + \cfrac{1}{2n + \cfrac{1}{2n+\ddots}} = [n, \overline{2n}].$$

Continued fractions for some other square roots are obtained from continued fractions with period 2 in the next example.

Example 5.8. *Some periodic continued fractions with period* 2**.** Replacing a with $2a$ in Theorem 5.9 yields

$$[\overline{2a, b}] = a + \sqrt{a^2 + 2a/b} = 2a + \cfrac{1}{b + \cfrac{1}{2a + \cfrac{1}{b+\ddots}}}$$

and subtracting a from both sides yields

$$\sqrt{a^2 + 2a/b} = a + \cfrac{1}{b + \cfrac{1}{2a + \cfrac{1}{b+\ddots}}} = [a, \overline{b, 2a}].$$

Some special cases are

$$\begin{aligned}
(a,b) = (n,n) : \sqrt{n^2+2} &= [n, \overline{n, 2n}],\\
(a,b) = (n,2) : \sqrt{n^2+n} &= [n, \overline{2, 2n}],\\
(a,b) = (n,1) : \sqrt{n^2+2} &= [n, \overline{1, 2n}],\\
(a,b) = (2n,4) : \sqrt{4n^2+n} &= [2n, \overline{4, 4n}], \text{ etc.}
\end{aligned}$$

□

5.9. Exercises

5.1 Use the result in Example 3.2 about integer squares modulo 3 to show that $\sqrt{3n-1}$ is irrational for $n \geq 1$. (Hint: First show that if $\sqrt{3n-1} = a/b$ for a and b relatively prime, then $a^2 + b^2 = 3nb^2$.)

5.2 (a) Show that a rational number to an irrational power may be irrational.

(b) Show that a rational number to an irrational power may be rational.

5.3 Prove that $\sqrt{n^2-1}$ is irrational for $n \geq 3$ an integer. [Hint: Modify proof 1 of Theorem 5.4.]

5.4 Prove that $\sqrt{n^2+1}$ is irrational for $n \geq 3$ an integer. [Hint: Modify the proof of Theorem 5.5.]

5.5 Show that (a) $\varphi^2 + (1/\varphi^2) = 3$ and (b) $\varphi^3 - (1/\varphi^3) = 4$. [Hints: (a) See Figure 5.12 and (b) construct a similar square with side length $\varphi + 1 + \varphi$.]

5.6 For the metallic ratios $\varphi_n = [\overline{n}]$ in Example 5.7 show that (a) $\varphi_n^2 = [1+n^2, \overline{1, n^2}]$ and (b) $\varphi_n^3 = \varphi_{n^3+3n} = [\overline{n^3+3n}]$.

5.7 In Example 5.4 we observed that

$$\cos\frac{\pi}{2^{n+1}} = \frac{1}{2}\sqrt{2+\sqrt{2+\sqrt{2+\cdots+\sqrt{2}}}}$$

with n nested square roots. Is there a similar expression for $\sin\frac{\pi}{2^{n+1}}$?

5.8 In Example 5.4 and Exercise 5.7 we encountered finitely many nested square roots. What meaning can be attached to infinitely many nested roots, such as

$$\sqrt{2+\sqrt{2+\sqrt{2+\cdots}}} \quad \text{or} \quad \sqrt{1+\sqrt{1+\sqrt{1+\cdots}}}?$$

[Hint: Examine the more general case $\sqrt{a+\sqrt{a+\sqrt{a+\cdots}}}$ with $a > 0$ first.]

CHAPTER 6

Fibonacci and Lucas Numbers

Learn the particular strength of the Fibonacci series,
a balanced spiraling outward of shapes,
whose golden numbers which describe dimensions
of sea shells, rams' horns, collections of petals
and generations of bees.
A formula to build your house on,
the proportion most pleasing to the human eye.

Judith Baumel

The elements of the sequence

$$\{F_n\}_{n=1}^{\infty} = \{1, 1, 2, 3, 5, 8, 13, 21, 34, 55, 89, \ldots\}$$

defined by $F_1 = F_2 = 1$ and $F_n = F_{n-1} + F_{n-2}$ for $n \geq 3$ are known as *Fibonacci numbers*, after the Italian mathematician Leonardo of Pisa (c. 1170–1240) or Leonardo Fibonacci (for *Filius Bonacci*, son of Bonacci) who first studied them (when convenient we set $F_0 = 0$). In [Koshy, 2001] Thomas Koshy describes the Fibonacci sequence and its twin, the Lucas sequence (see Section 6.4) as "the two shining stars in the vast array of integer sequences. They have fascinated both amateurs and professional mathematicians for centuries, and they continue to charm us with their beauty, their abundant applications, and their ubiquitous habit of occurring in totally surprising and unrelated places. They continue to be fertile ground for creative amateurs and mathematicians alike." The journal *The Fibonacci Quarterly*, first published in 1963, is devoted to the study of the properties of these sequences.

On the left below we see a 2013 postage stamp from the Principality of Liechtenstein. Across the bottom of the stamp are the first 19 Fibonacci numbers and the first three digits of $F_{20} = 6765$ (the 100 at the end is the stamp's denomination). On the right is a 1999 stamp from the Caribbean island nation of Dominica with a likeness of Leonardo.

Leonardo Fibonacci and his sequence

6.1. The Fibonacci sequence in art and nature

In this section we illustrate with two examples the use of the sequence in art and craft, and one example of its occurrence in nature.

A Fibonacci chimney

In the photographs below we see the Fibonacci sequence in neon on a chimney at the Turku Energia plant in Turku, Finland. It is a creation of the Italian artist Mario Merz (1925–2003) who has a similar installation on a chimney at the Center for International Light Art (formerly the Lindenbrauerei) in Unna, Germany, and on the dome of the Mole Antonelliana (now the Museo Nazionale del Cinema) in Turin, Italy.

(a) (b)

The Fibonacci chimney in Turku, Finland

There are a number of occurrences (or alleged occurrences) of the Fibonacci numbers in nature. The one we present in the following example concerns the ancestry of male honeybees.

Example 6.1. *The family tree of a male honeybee.* Male honeybees are born from unfertilized eggs (parthenogenesis), while female honeybees come from fertilized eggs. So males have mother but no father, while the females have both a mother and father. In Figure 6.1 we see the family tree of a male honeybee for six generations. The generations are numbered as follows: 1 is the male bee, 2 is its mother, 3 is its grandparents, etc. The numbers along the right edge of the tree are the number of bee ancestors in each generation.

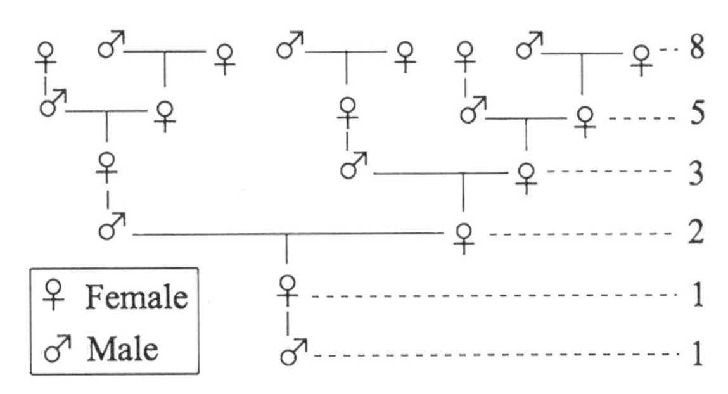

FIGURE 6.1. A portion of the family tree of a male honeybee

Let Q_n, D_n, and T_n denote, respectively, the number of female bees (queens), male bees (drones), and the total number of bees in generation n in the ancestry of the male in generation 1, where we have $Q_1 = 0$ and $D_1 = 1$. Then $T_n = Q_n + D_n$, $D_n = Q_{n-1}$ (since only queens have fathers), and $Q_n = Q_{n-1} + D_{n-1}$ (since both queens and drones have mothers) for $n \geq 1$. Hence $Q_n = Q_{n-1} + Q_{n-2}$ with $Q_1 = 0$ and $Q_2 = 1$, thus $Q_n = F_{n-1}$. Then $T_n = Q_n + Q_{n-1} = Q_{n+1} = F_n$, so that the number of bee ancestors in generation n is F_n. □

The Fibonacci sequence in textile design

The Fibonacci sequence appears to be popular with designers of textiles such as tartans, rugs, and quilts. In the tartan design below on the left, the widths of the stripes are elements of the sequence. Notice that the rectangles bordering the largest light gray square approximate the golden rectangle. In the center is a rug design with two copies of the sequence, one ascending, one descending. The quilt design on the right uses the same idea in both the vertical and horizontal directions.

A Fibonacci tartan, rug, and quilt.

6.2. Fibonacci parallelograms, triangles, and trapezoids

There exist many lovely identities relating the Fibonacci numbers. The first three we illustrate concern sums of Fibonacci numbers, specifically the sum of the first n, the sum of the first n with odd subscripts, and the sum of the first n with even subscripts:

$$\sum_{k=1}^{n} F_k = F_1 + F_2 + \cdots + F_n = F_{n+2} - 1; \tag{6.1}$$

$$\sum_{k=1}^{n} F_{2k-1} = F_1 + F_3 + \cdots + F_{2n-1} = F_{2n}; \tag{6.2}$$

$$\sum_{k=1}^{n} F_{2k} = F_2 + F_4 + \cdots + F_{2n} = F_{2n+1} - 1. \tag{6.3}$$

We illustrate these formulas with parallelograms, triangles, and trapezoids whose side lengths are Fibonacci numbers, as shown in Figure 6.2 [Walser, 2001]. Note that the sides of the parallelogram are consecutive Fibonacci numbers, and the top, the slanted sides, and the base of the trapezoid are three consecutive Fibonacci numbers.

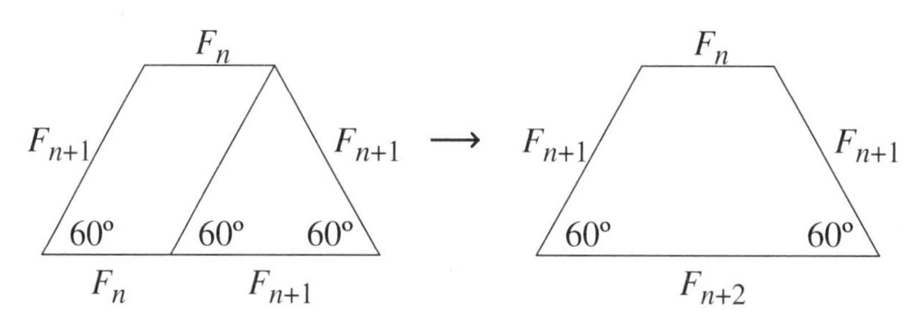

FIGURE 6.2

In Figure 6.3 we see an equilateral triangle subdivided into a number of small Δ and ∇ triangles with side lengths 1. The lengths of the left side and the base are the same, hence $1 + F_1 + F_3 + \cdots + F_n = F_{n+2}$, which is equivalent to (6.1).

In Figure 6.4 we recolor and relabel the previous figure, and gather together the light gray pieces into an equilateral triangle, and similarly for the dark gray pieces. In the light gray equilateral triangle, the lengths of the left side and the base are the same, hence $F_1 + F_3 + \cdots + F_{2n-1} = F_{2n}$, which is (6.2). Similarly, in the dark gray equilateral triangle, the lengths of the left side and the base are the same, hence $1 + F_2 + F_4 + \cdots + F_{2n} = F_{2n+1}$, which is equivalent to (6.3).

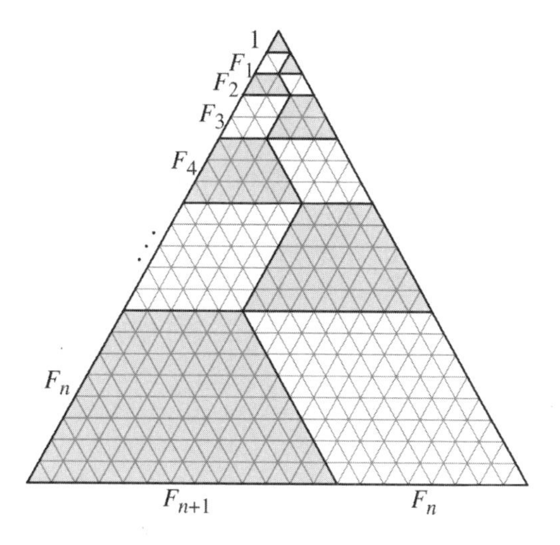

FIGURE 6.3

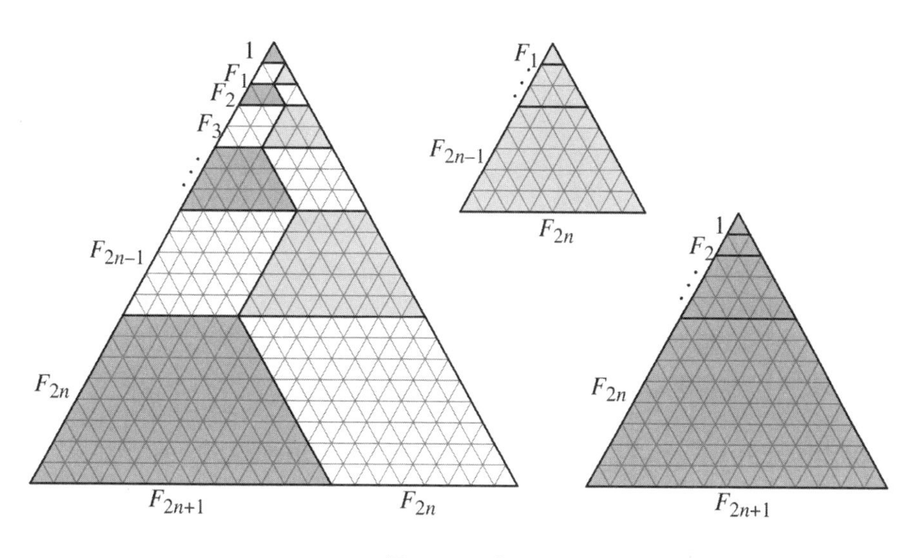

FIGURE 6.4

6.3. Fibonacci rectangles and squares

In this section we illustrate several identities involving products of Fibonacci numbers using areas of squares and rectangles. We apply the identities to show that consecutive Fibonacci numbers are relatively prime, construct Pythagorean triples, introduce the Fibonacci spiral, and establish some congruence results.

Example 6.2. An interesting Fibonacci identity results from comparing the square of one Fibonacci number to the product of its neighbors in the sequence, i.e., $2 \cdot 5 - 3^2 = +1, 3 \cdot 8 - 5^2 = -1, 5 \cdot 13 - 8^2 = +1$, and so on. This leads to an identity named for the Italian mathematician and astronomer Giovanni Domenico Cassini (1625–1712) who discovered it in 1680. □

Theorem 6.1. Cassini's identity. *For all* $n \geq 2$, $F_{n-1}F_{n+1} - F_n^2 = (-1)^n$.

Proof. Evaluating the area of the region in Figure 6.5(a) in two different ways yields

$$F_{n-1}F_{n+1} + F_{n-2}F_n = F_n^2 + F_{n-1}^2,$$

which is equivalent to $F_{n-1}F_{n+1} - F_n^2 = -\left(F_{n-2}F_n - F_{n-1}^2\right)$. Thus the terms in the sequence $\left\{F_{n-1}F_{n+1} - F_n^2\right\}_{n=2}^{\infty}$ have the same magnitude and alternate in sign. So we need only evaluate the base case $n = 2$: $F_1F_3 - F_2^2 = 1$ so $F_{n-1}F_{n+1} - F_n^2$ is $+1$ when n is even and -1 when n is odd, i.e., $F_{n-1}F_{n+1} - F_n^2 = (-1)^n$. □

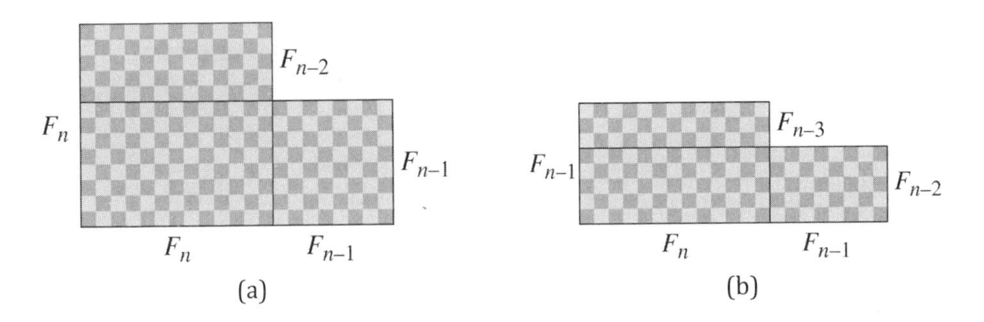

FIGURE 6.5

With Cassini's identity we can easily prove the following important property of the Fibonacci numbers.

Corollary 6.2. *Pairs of consecutive Fibonacci numbers are relatively prime, i.e.,* $\gcd(F_n, F_{n+1}) = 1$ *for* $n \geq 1$.

Proof. Suppose a prime p divides both F_n and F_{n+1}. Then by Cassini's identity, p divides $(-1)^n$, which is impossible. Hence $\gcd(F_n, F_{n+1}) = 1$. □

Example 6.3. A result similar to Cassini's identity holds for four consecutive Fibonacci numbers when we compare the product of the two outer numbers to that of the two inner numbers, i.e., $2 \cdot 8 - 3 \cdot 5 = +1$, $3 \cdot 13 - 5 \cdot 8 = -1$, $5 \cdot 21 - 8 \cdot 13 = +1$, etc. □

Theorem 6.3. *For all* $n \geq 3$, $F_{n-2}F_{n+1} - F_{n-1}F_n = (-1)^{n-1}$.

Proof. Computing the area of the region in Figure 6.5(b) in two ways yields $F_{n-2}F_{n+1} + F_{n-3}F_n = F_{n-1}F_n + F_{n-2}F_{n-1}$, equivalent to $F_{n-2}F_{n+1} - F_{n-1}F_n = -(F_{n-3}F_n - F_{n-2}F_{n-1})$. As in the proof of Theorem 6.1, the terms in the sequence $\{F_{n-2}F_{n+1} - F_{n-1}F_n\}_{n=3}^{\infty}$ have the same magnitude and alternate in sign. Evaluating the base case $n = 3$ yields $F_1F_4 - F_2F_3 = 1$, so $F_{n-2}F_{n+1} - F_{n-1}F_n$ is $+1$ when n is odd and -1 when n is even. □

Corollary 6.4. *Pairs of Fibonacci numbers whose subscripts differ by* 2 *are relatively prime, i.e.,* $\gcd(F_{n-1}, F_{n+1}) = 1$ *for* $n \geq 2$.

Proof. Replacing n by $n+1$ in Theorem 6.3 yields $F_{n-1}F_{n+2} - F_nF_{n+1} = (-1)^n$ for $n \geq 2$. If a prime p divides both F_{n-1} and F_{n+1}, then p divides $(-1)^n$, which is impossible. Hence $\gcd(F_{n-1}, F_{n+1}) = 1$. □

In the next example we illustrate a variety of identities for the square of a Fibonacci number.

Example 6.4. *Identities for* F_{n+1}^2. In Figure 6.6 we illustrate two different ways to partition a square with area F_{n+1}^2.

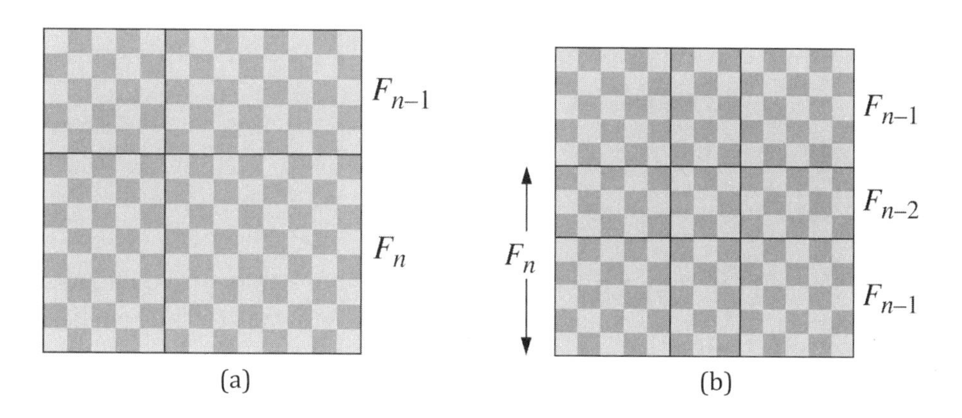

FIGURE 6.6

Figure 6.6a [Ollerton, 2008] illustrates the following five identities for the square of a Fibonacci number:

$$\begin{aligned}
F_{n+1}^2 &= 2F_{n+1}F_n - F_n^2 + F_{n-1}^2 \\
&= 2F_{n+1}F_{n-1} + F_n^2 - F_{n-1}^2 \\
&= F_n^2 + F_{n-1}^2 + 2F_nF_{n-1} \qquad (6.4) \\
&= F_{n+1}F_{n-1} + F_nF_{n-1} + F_n^2 \\
&= F_{n+1}F_n + F_nF_{n-1} + F_{n-1}^2. \qquad (6.5)
\end{aligned}$$

Similarly, Figure 6.6b [Brousseau, 1972] illustrates the following four identities:

$$\begin{aligned} F_{n+1}^2 &= 4F_nF_{n-1} + F_{n-2}^2 \\ &= 2F_n^2 + 2F_{n-1}^2 - F_{n-2}^2 \\ &= 4F_{n-1}^2 + 4F_{n-1}F_{n-2} + F_{n-2}^2 \\ &= 4F_n^2 - 4F_{n-1}F_{n-2} - 3F_{n-2}^2. \end{aligned}$$

Similar figures produce many other identities. □

Theorem 6.5. *For all positive integers n and k,*

$$F_{n+1}F_{k+1} + F_nF_k = F_{n+k+1}. \tag{6.6}$$

Proof. Evaluating the area of the region in Figure 6.7 in two ways yields

$$F_{n+1}F_{k+1} + F_nF_k = F_nF_{k+2} + F_{n-1}F_{k+1}. \tag{6.7}$$

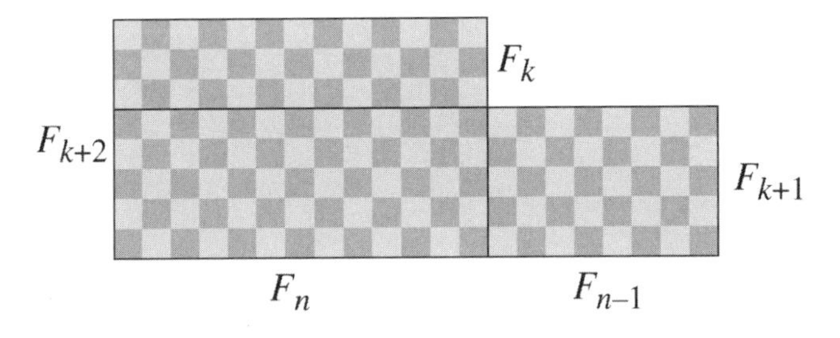

FIGURE 6.7

Observe that in the right-hand side of (6.7), n has been replaced by $n-1$ and k has been replaced by $k+1$, as compared to the left-hand side. Iterating (6.7) until n decreases to 2 yields

$$\begin{aligned} F_{n+1}F_{k+1} + F_nF_k &= F_nF_{k+2} + F_{n-1}F_{k+1} \\ &= F_{n-1}F_{k+3} + F_{n-2}F_{k+2} \\ &= F_{n-2}F_{k+4} + F_{n-3}F_{k+3} \\ &\vdots \\ &= F_2F_{n+k} + F_1F_{n+k-1} = F_{n+k+1}, \end{aligned}$$

where in the final step we use $F_1 = F_2 = 1$ and $F_{n+k} + F_{n+k-1} = F_{n+k+1}$. □

Corollary 6.6. *Each Fibonacci number is either the sum or the difference of squares of two smaller Fibonacci numbers, i.e., for n positive,* $F_{n+1}^2 + F_n^2 = F_{2n+1}$ *and* $F_{n+1}^2 - F_{n-1}^2 = F_{2n}$.

Proof. Setting $k = n$ in (6.6) yields $F_{n+1}^2 + F_n^2 = F_{2n+1}$. Identity (6.5) derived from Figure 6.6a is equivalent to $F_{n+1}^2 - F_{n-1}^2 = F_{n+1}F_n + F_nF_{n-1}$, and setting $k = n - 1$ in (6.6) yields $F_{n+1}F_n + F_nF_{n-1} = F_{2n}$, so that $F_{n+1}^2 - F_{n-1}^2 = F_{2n}$. □

The relationship $F_{n+1}^2 + F_n^2 = F_{2n+1}$ for the Fibonacci sequence mirrors the relationship $T_k^2 + T_{k-1}^2 = T_{k^2}$ for the sequence of triangular numbers encountered in Example 1.5: The sum of the squares of two consecutive members of the sequence is another member of the sequence.

Corollary 6.7. *If n divides m, then F_n divides F_m.*

Proof. Setting $k = n - 1$ in (6.6) yields $F_{n+1}F_n + F_nF_{n-1} = F_{2n}$, so that F_n divides F_{2n}. Setting $k = 2n - 1$ in (6.6) yields $F_{n+1}F_{2n} + F_nF_{2n-1} = F_{3n}$, so that F_n divides F_{3n}. Continuing in this fashion shows that F_n divides F_{kn} for any positive integer k. □

As a consequence of Corollary 6.7, $F_3 = 2$ implies that every third Fibonacci number is even (i.e., $F_{3k} \equiv 0 \pmod 2$ for $k \geq 1$); $F_4 = 3$ implies that every fourth Fibonacci number is a multiple of 3 (i.e., $F_{4k} \equiv 0 \pmod 3$ for $k \geq 1$); $F_5 = 5$ implies that every fifth Fibonacci number is a multiple of 5 (i.e., $F_{5k} \equiv 0 \pmod 5$ for $k \geq 1$); $F_6 = 8$ implies that every sixth Fibonacci number is a multiple of 8 (i.e., $F_{6k} \equiv 0 \pmod 8$ for $k \geq 1$); and so on.

Example 6.5. *Fibonacci Pythagorean triples.* Since F_k and F_{k+1} for $k \geq 2$ are relatively prime, we can use them to generate PTs (which will be primitive when F_k and F_{k+1} have opposite parity). Setting $(m, n) = (F_k, F_{k+1})$ for $k \geq 2$ in Euclid's formula yields $a = F_{k+1}^2 - F_k^2 = F_{k-1}F_{k+2}$, $b = 2F_kF_{k+1}$, and $c = F_k^2 + F_{k+1}^2 = F_{2k+1}$. Thus every Fibonacci number with an odd subscript, beginning with $F_5 = 5$, is the hypotenuse of some PT, and the area of the PT is $F_{k-1}F_kF_{k+1}F_{k+2}$, the product of four consecutive Fibonacci numbers.

Similarly, F_{k-1} and F_{k+1} are relatively prime, so setting $(m, n) = (F_{k-1}, F_{k+1})$ in Euclid's formula yields $a = F_{k+1}^2 - F_{k-1}^2 = F_{2k}$, $b = 2F_{k-1}F_{k+1}$, and $c = F_{k-1}^2 + F_{k+1}^2$. Thus every Fibonacci number with an even subscript, beginning with $F_4 = 3$, is the leg of some PT. These results are not surprising, since every number greater than 2 appears in some PT. □

Example 6.6. *Sums of Fibonacci squares and the Fibonacci spiral.* A little experimentation leads to the hypothesis that the sum of the first n

Fibonacci squares is a product of consecutive Fibonacci numbers:

$$
\begin{aligned}
1^2 + 1^2 &= 2 = 1 \cdot 2, \\
1^2 + 1^2 + 2^2 &= 6 = 2 \cdot 3, \\
1^2 + 1^2 + 2^2 + 3^2 &= 15 = 3 \cdot 5, \\
1^2 + 1^2 + 2^2 + 3^2 + 5^2 &= 40 = 5 \cdot 8, \\
1^2 + 1^2 + 2^2 + 3^2 + 5^2 + 8^2 &= 104 = 8 \cdot 13, \text{ etc.}
\end{aligned}
$$

Theorem 6.8. $F_1^2 + F_2^2 + \cdots + F_n^2 = F_n F_{n+1}$.

Proof. See Figure 6.8a [Brousseau, 1972] . □

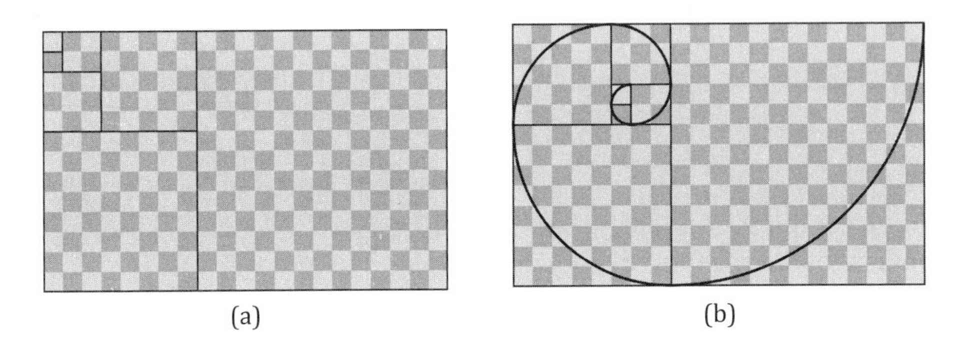

(a) (b)

FIGURE 6.8

If we rearrange some of the squares in Figure 6.8a and draw a quarter circle in each one, we create a portion of the beautiful *Fibonacci spiral*, as illustrated in Figure 6.8b. This spiral appears on the 1987 Swiss postage stamp, a 10 litas Lithuanian gold coin minted in 2007, and in a floor tiling in the Julian Science and Mathematics Center at DePauw University in Greencastle, Indiana, as illustrated in Figure 6.9. □

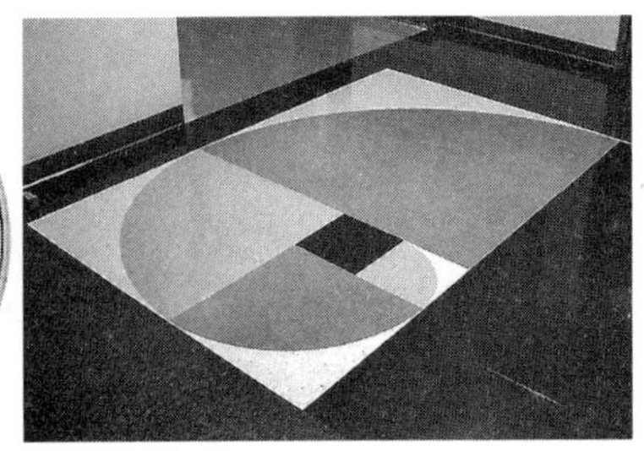

FIGURE 6.9

Example 6.7. *Sums of Fibonacci rectangles.* If we replace the Fibonacci squares in the preceding example with rectangles, we obtain two more identities.

Theorem 6.9.

$$\text{(a)}\quad F_1F_2 + F_2F_3 + \cdots + F_{2n-2}F_{2n-1} = F_{2n-1}^2 - 1,$$

$$\text{(b)}\quad F_1F_2 + F_2F_3 + \cdots + F_{2n-1}F_{2n} = F_{2n}^2.$$

See Figure 6.10 where we illustrate the identities for $n = 4$. □

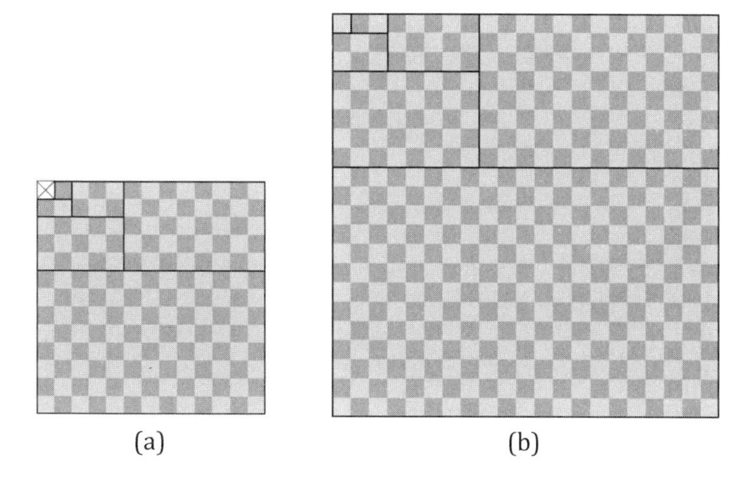

(a) (b)

FIGURE 6.10

6.4. Diagonal sums in Pascal's triangle

In Figure 6.11 we see a portion of Pascal's triangle, presented left-justified so that the binomial coefficient $\binom{n}{k}$ appears in the nth row and kth column for both n and k in $\{0, 1, 2, 3, \ldots\}$. The gray arrows indicate sums of the elements on the rising diagonals, which appear to yield Fibonacci numbers. We also indicate the use of *Pascal's formula* (also known as *Pascal's rule* or *Pascal's identity*) by the boxed L-shaped region for adding two consecutive elements in one row to obtain an element in the next row, e.g., $15 + 6 = 21$.

Let D_n denote the sum of the elements in the nth diagonal for $n \geq 0$. Then $D_0 = \binom{0}{0} = 1$, $D_1 = \binom{1}{0} = 1$, and

$$D_n = \binom{n}{0} + \binom{n-1}{1} + \cdots = \sum_{k=0}^{\lfloor n/2 \rfloor} \binom{n-k}{k}$$

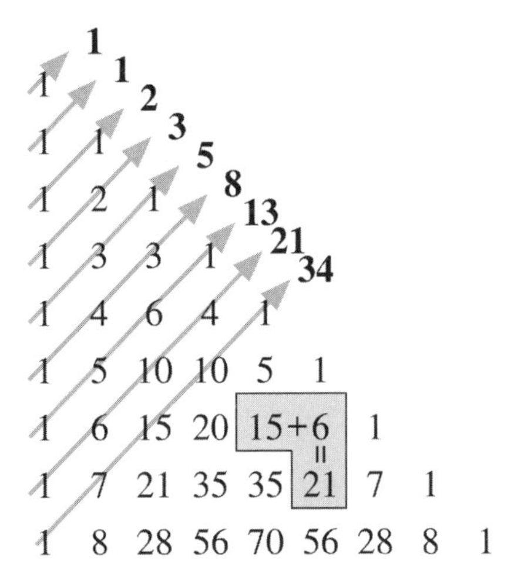

FIGURE 6.11

for $n \geq 2$. If we can show that $D_n + D_{n+1} = D_{n+2}$ for $n \geq 0$, then we can conclude that $D_n = F_{n+1}$ for $n \geq 0$.

In Figure 6.12a we use Pascal's formula to illustrate $D_n + D_{n+1} = D_{n+2}$ for n odd, e.g., for $n = 5$; and in Figure 6.12b for n even, e.g., for $n = 6$. We have used the fact that $\binom{n}{k}$ is defined to be 0 whenever $k < 0$ or $k > n$, as indicated by the gray 0's.

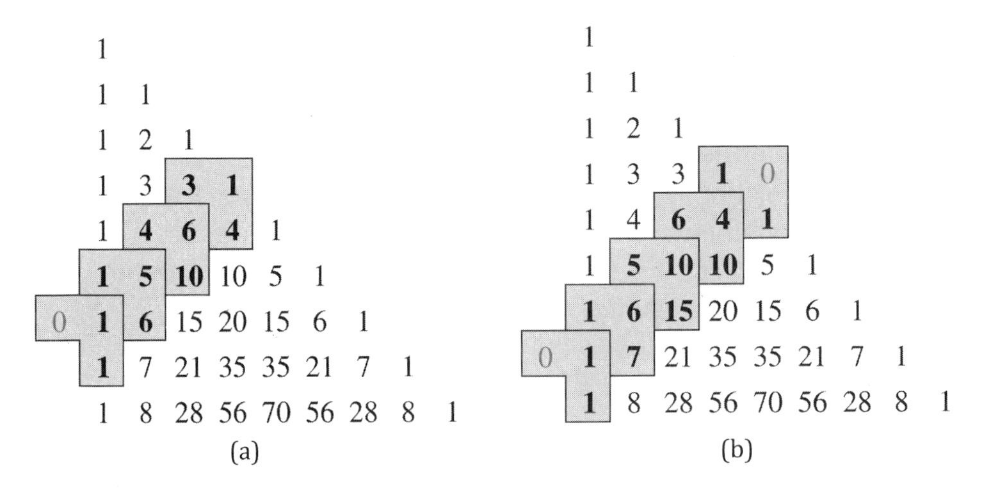

FIGURE 6.12

6.5. Lucas numbers

A sequence closely related to the Fibonacci sequence is the sequence of *Lucas numbers,*

$$\{L_n\}_{n=1}^{\infty} = \{1, 3, 4, 7, 11, 18, 29, 47, 76, 123, 199, ...\}$$

defined by $L_1 = 1, L_2 = 3$, and $L_n = L_{n-1} + L_{n-2}$ for $n \geq 3$. When convenient we set $L_0 = 2$. This sequence is named for the French mathematician François Edouard Anatole Lucas (1842–1891), who made many contributions to the study of the Fibonacci numbers and is responsible for giving that sequence its name. Since the Lucas sequence has the same recursion formula as the Fibonacci sequence, many of its identities are similar. Here are some examples for the Lucas numbers, and for both Lucas and Fibonacci numbers.

Example 6.8. The Lucas analogue of (6.1) is

$$\sum_{k=1}^{n} L_k = L_1 + L_2 + \cdots + L_n = L_{n+2} - 3. \tag{6.8}$$

The visual proof using Fibonacci trapezoids, as in Figure 6.3, is somewhat cumbersome for this identity. But converting the sum into a telescoping sum succeeds nicely:

$$\sum_{k=1}^{n} L_k = \sum_{k=1}^{n} (L_{k+2} - L_{k+1}) = L_{n+2} - L_2 = L_{n+2} - 3. \qquad \square$$

Example 6.9. *Cassini's identity for the Lucas numbers* is given by $L_{n-1}L_{n+1} - L_n^2 = 5(-1)^{n+1}$ for $n \geq 2$. The proof of this identity is similar to the proof of Theorem 6.1. Relabeling the image in Figure 6.8a with Lucas numbers yields $L_{n-1}L_{n+1} + L_{n-2}L_n = L_n^2 + L_{n-1}^2$, which is equivalent to $L_{n-1}L_{n+1} - L_n^2 = -\left(L_{n-2}L_n - L_{n-1}^2\right)$. Thus the terms in the sequence $\left\{L_{n-1}L_{n+1} - L_n^2\right\}_{n=2}^{\infty}$ have the same magnitude and alternate in sign. So we need only evaluate the base case $n = 2$: $L_1L_3 - L_2^2 = 1 \cdot 4 - 3^2 = -5$ so $L_{n-1}L_{n+1} - L_n^2$ is -5 when n is even and $+5$ when n is odd, i.e., $L_{n-1}L_{n+1} - L_n^2 = 5(-1)^{n+1}$. $\square$

In Exercise 6.4 you can prove the Lucas analogues of identities (6.2) and (6.3); and in Exercise 6.6 you can derive the Lucas analogue of the identity in Theorem 6.5.

There are many lovely identities involving both the Fibonacci numbers and the Lucas numbers. Here is one example.

Theorem 6.10. *For all positive integers n and k,*

$$F_{k+1}L_{n+1} + F_kL_n = L_{n+k+1}.$$

Proof. Evaluating the area of the region in Figure 6.13a in two ways yields

$$F_{k+1}L_{n+1} + F_kL_n = F_kL_{n+2} + F_{k-1}L_{n+1}. \tag{6.9}$$

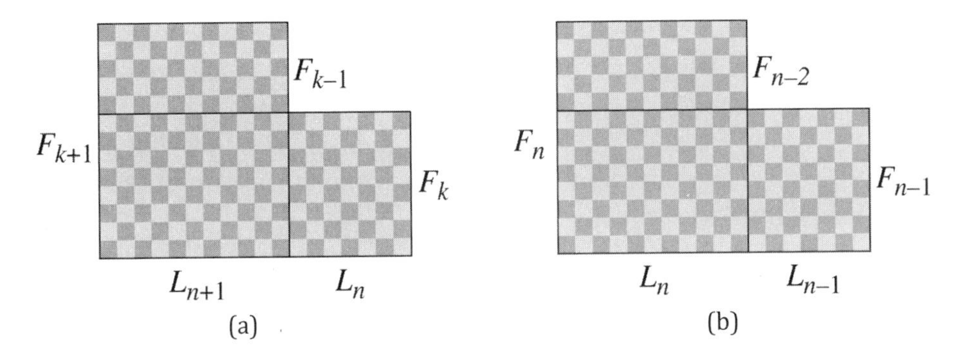

FIGURE 6.13

Observe that in the right-hand side of (6.9), k has been replaced by $k-1$ and n has been replaced by $n+1$, as compared to the left-hand side. Iterating (6.9) until k decreases to 2 yields

$$\begin{aligned}
F_{k+1}L_{n+1} + F_kL_n &= F_kL_{n+2} + F_{k-1}L_{n+1} \\
&= F_{k-1}L_{n+3} + F_{k-2}L_{n+2} \\
&= F_{k-2}L_{n+4} + F_{k-3}L_{n+3} \\
&\vdots \\
&= F_2L_{n+k} + F_1L_{n+k-1} = L_{n+k+1},
\end{aligned}$$

where in the final step we use $F_1 = F_2 = 1$ and $L_{n+k} + L_{n+k-1} = L_{n+k+1}$. □

The image in Figure 6.13b, a relabeling of Figure 6.13a, can be used to prove the following theorem, a Cassini-like identity for products of Fibonacci and Lucas numbers.

Theorem 6.11. *For all $n \geq 2$, $F_{n-1}L_{n+1} - F_nL_n = (-1)^n$.*

Compare this identity to those in Theorem 6.1 [$F_{n-1}F_{n+1} - F_n^2 = (-1)^n$] and Example 6.9 [$L_{n-1}L_{n+1} - L_n^2 = 5(-1)^{n+1}$].

Proof. Evaluating the area of the region in Figure 6.13b in two different ways yields

$$F_{n-1}L_{n+1} + F_{n-2}L_n = F_nL_n + F_{n-1}L_{n-1},$$

which is equivalent to $F_{n-1}L_{n+1} - F_nL_n = -(F_{n-2}L_n - F_{n-1}L_{n-1})$. Thus the terms in the sequence $\{F_{n-1}L_{n+1} - F_nL_n\}_{n=2}^{\infty}$ have the same magnitude and alternate in sign. So we need only evaluate the base case $n = 2$: $F_1L_3 - F_2L_2 = 1$ so $F_{n-1}L_{n+1} - F_nL_n$ is $+1$ when n is even and -1 when n is odd, i.e., $F_{n-1}L_{n+1} - F_nL_n = (-1)^n$. □

With the identity in Theorem 6.11 we can easily prove the following property of the Lucas numbers, analogous to Corollary 6.2 for the Fibonacci numbers.

Corollary 6.12. *Pairs of consecutive Lucas numbers are relatively prime, i.e.,* $\gcd(L_n, L_{n+1}) = 1$ *for* $n \geq 1$.

Proof. Suppose a prime p divides both L_n and L_{n+1}. Then by Theorem 6.11, p divides $(-1)^n$, which is impossible. Hence $\gcd(L_n, L_{n+1}) = 1$. □

6.6. The Pell equations $x^2 - 5y^2 = \pm 4$ and Binet's formula

In Corollary 6.6 we established the identity $F_{n+1}^2 + F_n^2 = F_{2n+1}$, and the Lucas version $L_{n+1}^2 + L_n^2 = 5F_{2n+1}$ follows from setting $k = n$ into the identity in Exercise 6.6 at the end of this chapter. Combining these yields $L_{n+1}^2 + L_n^2 = 5F_{n+1}^2 + 5F_n^2$ or, equivalently, $L_{n+1}^2 - 5F_{n+1}^2 = -\left(L_n^2 - 5F_n^2\right)$. Thus the terms in the sequence $\left\{L_n^2 - 5F_n^2\right\}_{n=1}^{\infty}$ have the same magnitude and alternate in sign. Again we need only evaluate the base case $n = 1$: $L_1^2 - 5F_1^2 = -4$; hence $L_n^2 - 5F_n^2 = 4(-1)^n$ for $n \geq 1$. This identity was discovered in 1950 by P. Schub [Schub, 1950].

Thus the pairs (L_n, F_n) are solutions to the Pell equations $x^2 - 5y^2 = \pm 4$. We now solve these Pell equations using the methods presented in Chapter 3 to illustrate how the pairs (L_n, F_n) arise as solutions, and proceed to find explicit formulas for the Fibonacci and Lucas numbers.

In Section 5.2 we established the identity $x^2 - 2y^2 = 2(x+y)^2 - (x+2y)^2$ using Figure 5.4. This is the $d = 2$ case for the more general identity $(d-1)(x^2 - dy^2) = d(x+y)^2 - (x+dy)^2$, which holds for all positive integers x, y, and d. When $d = 5$ we have $4(x^2 - 5y^2) = 5(x+y)^2 - (x+5y)^2$, which we illustrate in Figure 6.14 in the form $(x+5y)^2 + 4x^2 = 20y^2 + 5(x+y)^2$.

The identity $4(x^2 - 5y^2) = 5(x+y)^2 - (x+5y)^2$ is equivalent to

$$x^2 - 5y^2 = 5\left(\frac{x+y}{2}\right)^2 - \left(\frac{x+5y}{2}\right)^2.$$

This establishes the following theorem, an analogue of Theorem 5.2.

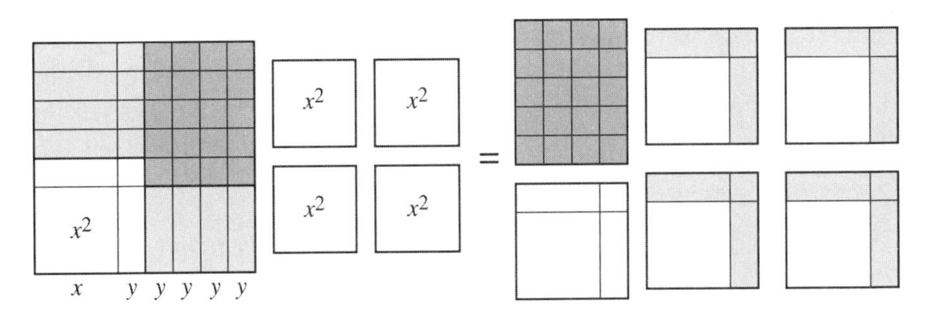

FIGURE 6.14

Theorem 6.13. *Let x and y be positive integers. Then*

$$\left(\frac{x+5y}{2}\right)^2 - 5\left(\frac{x+y}{2}\right)^2 = \pm 4 \text{ if and only if } x^2 - 5y^2 = \mp 4. \tag{6.10}$$

Note that solutions to $x^2 - 5y^2 = \pm 4$ must have x and y both even or both odd, hence $(x+y)/2$ and $(x+5y)/2$ are integers. Set

$$x_{n+1} = \frac{1}{2}(x_n + 5y_n) \quad \text{and} \quad y_{n+1} = \frac{1}{2}(y_n + x_n) \tag{6.11}$$

in (6.10) yielding $x_{n+1}^2 - 5y_{n+1}^2 = \pm 4$ if and only if $x_n^2 - 5y_n^2 = \mp 4$. Setting $(x_1, y_1) = (1, 1)$ yields $x_1^2 - 5y_1^2 = -4$, hence $x_n^2 - 5y_n^2 = 4(-1)^n$. In Table 6.1 we have the solutions (x_n, y_n) to $x_n^2 - 5y_n^2 = 4(-1)^n$ generated by (6.11) for n from 1 to 9.

TABLE 6.1

n	1	2	3	4	5	6	7	8	9
x_n	1	3	4	7	11	18	29	47	76
y_n	1	1	2	3	5	8	13	21	34

It appears that $x_n = L_n$ and $y_n = F_n$, but to verify this we must show that $x_{n+1} = x_n + x_{n-1}$ and $y_{n+1} = y_n + y_{n-1}$ for $n \geq 2$. From (6.11) we have

$$\begin{aligned} x_{n+1} &= \frac{1}{2}(x_n + 5y_n) \\ &= \frac{1}{2}x_n + \frac{5}{2}\left(\frac{1}{2}x_{n-1} + \frac{1}{2}y_{n-1}\right) \\ &= \frac{1}{2}x_n + \frac{5}{4}x_{n-1} + \frac{1}{4}(2x_n - x_{n-1}) \\ &= x_n + x_{n-1} \end{aligned}$$

and

$$\begin{aligned} y_{n+1} &= \frac{1}{2}(x_n + y_n) \\ &= \frac{1}{2}y_n + \frac{1}{2}\left(\frac{5}{2}y_{n-1} + \frac{1}{2}x_{n-1}\right) \\ &= \frac{1}{2}y_n + \frac{5}{4}y_{n-1} + \frac{1}{4}(2y_n - y_{n-1}) \\ &= y_n + y_{n-1}. \end{aligned}$$

Since $x_1 = 1, x_2 = 3$, and $y_1 = y_2 = 1$, we have $x_n = L_n$ and $y_n = F_n$ for $n \geq 1$, and (6.11) yields the identities

$$L_{n+1} = \frac{1}{2}(L_n + 5F_n) \text{ and } F_{n+1} = \frac{1}{2}(F_n + L_n). \tag{6.12}$$

Example 6.10. *A square root test for Fibonacci-ness.* Is 25,075 a Fibonacci number? Is 75,025 a Fibonacci number? If y is a Fibonacci number, then by the identity $L_n^2 - 5F_n^2 = 4(-1)^n$, either $\sqrt{5y^2 - 4}$ or $\sqrt{5y^2 + 4}$ is an integer (in fact, a Lucas number). When $y = 25,075$, neither $\sqrt{5(25,075)^2 - 4}$ nor $\sqrt{5(25,075)^2 + 4}$ is an integer, so 25,075 is not a Fibonacci number. But when $y = 75,025$, $\sqrt{5(75,025)^2 - 4} = 167,761$, so 75,025 is a Fibonacci number (it is F_{25} and $167,761 = L_{25}$). □

Example 6.11. *Binet's formula.* We now find explicit expressions for F_n and L_n. Evaluating the generating functions $G_F(t) = \sum_{n=0}^{\infty} F_n t^n$ and $G_L(t) = \sum_{n=0}^{\infty} L_n t^n$ for the sequences $\{F_n\}$ and $\{L_n\}$ using $F_{n+2} = F_{n+1} + F_n$ and $L_{n+2} = L_{n+1} + L_n$ for $n \geq 0$ yields $G_F(t) = t/(1 - t - t^2)$ and $G_L(t) = (2 - t)/(1 - t - t^2)$, from which we obtain

$$F_n = \frac{1}{\sqrt{5}}\left[\left(\frac{1+\sqrt{5}}{2}\right)^n - \left(\frac{1-\sqrt{5}}{2}\right)^n\right]$$

and

$$L_n = \left(\frac{1+\sqrt{5}}{2}\right)^n + \left(\frac{1-\sqrt{5}}{2}\right)^n$$

using partial fractions and geometric series. The formula for F_n is known as *Binet's formula,* after the French mathematician Jacques Philippe Marie Binet (1786–1856) who discovered the formula in 1843. However, it was known a century earlier to another French mathematician, Abraham de Moivre (1667–1754).

The fraction $\frac{1+\sqrt{5}}{2}$ is, of course, the golden ratio φ, and $\frac{1-\sqrt{5}}{2} = -\frac{1}{\varphi}$, so we have simple expressions for F_n and L_n in terms of the golden ratio:

$$F_n = \frac{1}{\sqrt{5}}\left[\varphi^n - (-\varphi)^{-n}\right] \text{ and } L_n = \varphi^n + (-\varphi)^{-n}$$

or, since $1 - \varphi = -\frac{1}{\varphi}$,

$$F_n = \frac{1}{\sqrt{5}}\left[\varphi^n - (1-\varphi)^n\right] \text{ and } L_n = \varphi^n + (1-\varphi)^n. \qquad \square$$

Example 6.12. *Rational approximations to* $\sqrt{5}$. Since Fibonacci and Lucas numbers are solutions to the Pell equations $x^2 - 5y^2 = \pm 4$, their ratios approximate $\sqrt{5}$. Dividing $L_n^2 - 5F_n^2 = 4(-1)^n$ by F_n^2 yields

$$\left(\frac{L_n}{F_n}\right)^2 - 5 = (-1)^n \frac{4}{F_n^2}.$$

Thus $\lim_{n\to\infty} L_n/F_n = \sqrt{5}$, and L_n/F_n approximates $\sqrt{5}$ (larger than $\sqrt{5}$ when n is even, smaller when n is odd) with an absolute error less than $4/3F_n^2$. This error bound follows from

$$\frac{4}{F_n^2} = \left(\frac{L_n}{F_n} + \sqrt{5}\right)\left|\frac{L_n}{F_n} - \sqrt{5}\right| > 3\left|\frac{L_n}{F_n} - \sqrt{5}\right|$$

since $\frac{L_n}{F_n} + \sqrt{5} \geq 1 + \sqrt{5} > 3$. $\square$

Example 6.13. *Limits of* F_{n+1}/F_n *and* L_{n+1}/L_n. There are various ways to examine the behavior of F_{n+1}/F_n and L_{n+1}/L_n as n increases without bound. The values of the ratio F_{n+1}/F_n for various values of n can be seen in Figures 6.8 and 6.10, where the rectangles with dimensions $F_{n+1} \times F_n$ appear to be approximately "golden" (see Figures 5.10 and 5.11).

A simple way to evaluate $\lim_{n\to\infty} F_{n+1}/F_n$ utilizes the second identity in (6.12) and Example 6.12:

$$\lim_{n\to\infty} \frac{F_{n+1}}{F_n} = \lim_{n\to\infty} \frac{1}{2}\left(1 + \frac{L_n}{F_n}\right) = \frac{1}{2}(1+\sqrt{5}) = \varphi$$

and similarly $\lim_{n\to\infty} L_{n+1}/L_n = \varphi$ from the first identity in (6.12). $\square$

For a comprehensive overview of Fibonacci and Lucas numbers and their properties, see [Koshy, 2001].

6.7. Exercises

6.1 Use illustrations similar to Figures 6.8 and 6.10 to show that

$$F_1F_3 + F_2F_4 + \cdots + F_{2n}F_{2n+2} = F_2^2 + F_3^2 + F_4^2 + \cdots + F_{2n+1}^2$$
$$= F_{2n+1}F_{2n+2} - 1.$$

[Hint: Partition an $F_{2n+1} \times F_{2n+2}$ rectangle (with one square missing) in two different ways.]

6.2 In Example 6.4 we partitioned a square to illustrate (6.4), the identity $F_{n+1}^2 = F_n^2 + F_{n-1}^2 + 2F_nF_{n-1}$. A cubic version is $F_{n+1}^3 = F_n^3 + F_{n-1}^3 + 3F_{n+1}F_nF_{n-1}$. Can you partition a cube to illustrate this?

6.3 Show that the Lucas sequence can be obtained from the Fibonacci sequence and vice-versa: $F_{n-1} + F_{n+1} = L_n$ and $L_{n-1} + L_{n+1} = 5F_n$.

6.4 Prove the Lucas analogues of identities (6.2) and (6.3), i.e.,
(a) $\sum_{k=1}^{n} L_{2k-1} = L_{2n} - 2$ and
(b) $\sum_{k=1}^{n} L_{2k} = L_{2n+1} - 1$.

6.5 Show that (a) $F_nL_n = F_{2n}$ and (b) $F_kL_n + F_nL_k = 2F_{n+k}$.

6.6 Prove the Lucas analogue of Theorem 6.5: $L_{n+1}L_{k+1} + L_nL_k = 5F_{n+k+1}$.

6.7 Prove the Lucas analogue of Theorem 6.8: $L_1^2 + L_2^2 + \cdots + L_n^2 = L_nL_{n+1} - 2$.

6.8 The numbers φ^n and $(-\varphi)^{-n}$ appear in formulas for F_n and L_n in Example 6.11. When $n = 1$ these numbers are the roots of the quadratic equation $x^2 - x - 1 = 0$. Find a quadratic equation with roots φ^n and $(-\varphi)^{-n}$.

6.9 We showed that $F_{3k} \equiv 0 \pmod{2}$ for $k \geq 1$ as a consequence of Corollary 6.7. Now show that $L_{3k} \equiv 0 \pmod{2}$ for $k \geq 1$.

6.10 Show that (a) the arithmetic mean of F_n and L_n is F_{n+1}, and (b) the harmonic mean of F_n and L_n is F_{2n}/F_{n+1}.

CHAPTER 7

Perfect Numbers

Perfect numbers, like perfect men, are very rare.

René Descartes

Man ever seeks perfection but inevitably it eludes him. He has sought perfect numbers through the ages and has found a very few.

Albert H. Beiler

A *perfect number* is a positive integer n that is equal to the sum of its *proper divisors* (those divisors less than n). For example, 6 is perfect since $6 = 1 + 2 + 3$, and 28 is perfect since $28 = 1 + 2 + 4 + 7 + 14$. Perfect numbers have been studied since the time of the early Greek mathematicians. Perfect numbers appear in Definition 22 in Book VII of Euclid's *Elements* [Heath, 1956] as "*A perfect number is that which is equal to its own parts.*" The Greeks were aware of the first four perfect numbers, 6, 28, 496, and 8128, and in the Middle Ages mathematicians found the next three: 33,550,336, 8,589,869,056, and 137,438,691,328. Two of the oldest unanswered questions in mathematics are (1) do infinitely many perfect numbers exist, and (2) are there any odd perfect numbers?

7.1. Euclid's formula

In addition to the definition mentioned above, Euclid provides a formula for perfect numbers in Proposition 36 in Book IX of the *Elements*: "*If as many numbers as we please beginning from a unit be set out in double proportion, until the sum of all becomes prime, and if the sum is multiplied into the last make some number, the product will be perfect.*" The expression "double proportion" refers to a geometric progression with common ratio 2. For example, $1+2+4 = 7$ and 7 is prime, so $4 \cdot 7 = 28$ is perfect. The numbers constructed by this formula are necessarily even.

Since $1 + 2 + 4 + \cdots + 2^{k-1} = 2^k - 1$ (a formula known to the early Greeks), Euclid's formula states that if $2^k - 1$ is prime, then $2^{k-1}(2^k - 1)$ is perfect. But $2^k - 1$ prime implies that k is prime, so we have

Theorem 7.1. Euclid's formula for even perfect numbers. *If p and $q = 2^p - 1$ are prime, then $N_p = 2^{p-1}q$ is perfect.*

Proof. The proper divisors of N_p are $1, 2, 4, \ldots, 2^{p-1}, q, 2q, 4q, \ldots, 2^{p-2}q$. Figure 7.1 [Goldberg] shows that rectangles with areas equal to the proper divisors sum to $N_p = 2^{p-1}q$. □

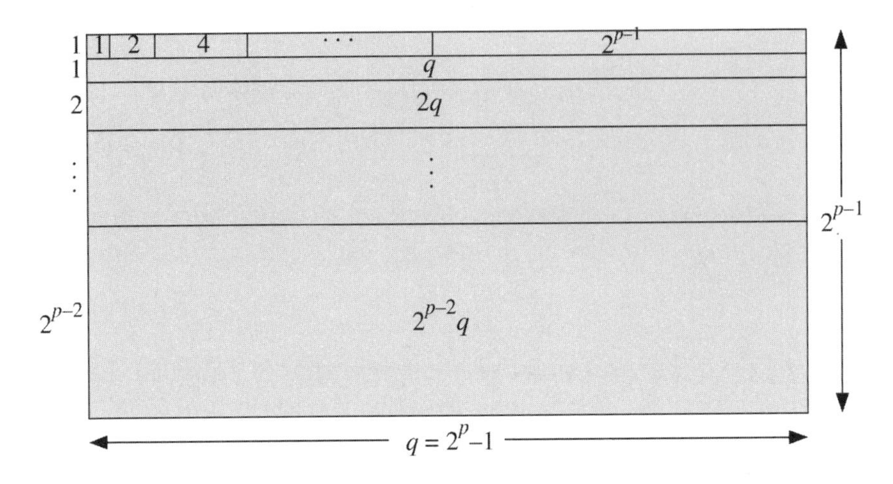

FIGURE 7.1

Nearly 2000 years after Euclid, the Swiss mathematician Leonard Euler (1807–1783) proved that every even perfect number has the form expressed by Euclid.

Leonhard Euler, Swiss postage stamp, 2007

Mersenne primes

Mersenne primes, named for the French mathematician and theologian Marin Mersenne (1588–1648), are primes of the form $q = 2^p - 1$. These are precisely the primes q used to construct even perfect numbers in Theorem 7.1. Thus the search for even perfect numbers is the search for

Mersenne primes. As of 2018, 50 Mersenne primes are known, the largest $2^{77,232,917} - 1$ has 23,249,425 digits. The 2004 Liechtenstein postage stamp shown below celebrates the 39th Mersenne prime, discovered in 2001. The 15 largest known Mersenne primes were discovered by the Great Internet Mersenne Prime Search (GIMPS) at www.mersenne.org.

Marin Mersenne and the 39th Mersenne prime

7.2. Even perfect numbers and geometric progressions

Theorem 7.2. *The even perfect number N_p is the sum of p terms of a geometric progression with first term 2^{p-1} and common ratio 2, i.e., $N_p = 2^{p-1} + 2^p + \cdots + 2^{2p-2}$.*

For example, when $p = 5$ we have $N_5 = 496 = 16+32+64+128+256$.

Proof. We slice the rectangle representing N_p in Figure 7.1 vertically, as shown in Figure 7.2, to form p rectangles whose areas are the terms of the geometric progression. □

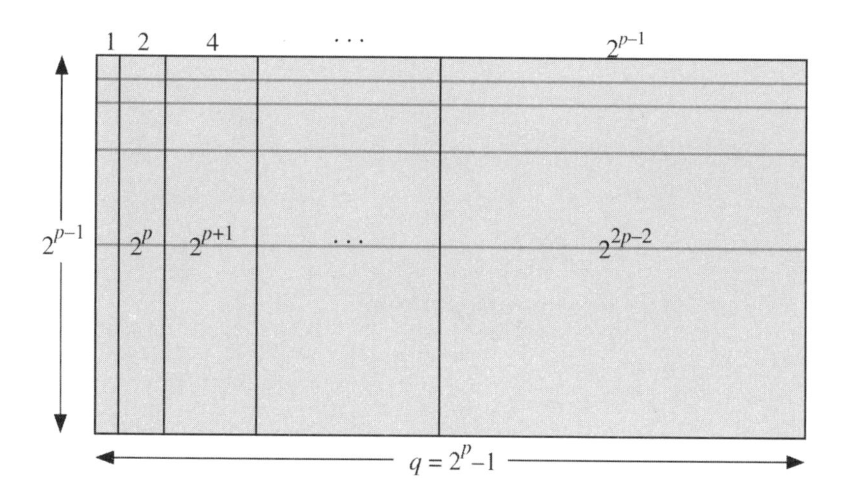

FIGURE 7.2

The algebraic proof is equally simple:

$$N_p = 2^{p-1}\left(2^p - 1\right) = 2^{p-1}\left(1 + 2 + 4 + \cdots + 2^{p-1}\right).$$

7.3. Even perfect numbers and triangular numbers

Each even perfect number is also the sum of an arithmetic progression; e.g.,

$$\begin{aligned} N_2 &= 1 + 2 + 3, \\ N_3 &= 1 + 2 + 3 + 4 + 5 + 6 + 7, \\ N_5 &= 1 + 2 + 3 + \cdots + 30 + 31, \text{ etc.} \end{aligned}$$

Each sum is a triangular number of the form $N_p = T_{2^p-1}$, as shown in Figure 7.3 (for $p = 5$, $N_5 = 16 \cdot 31$). Here we represent N_p as a $2^{p-1} \times (2^p - 1)$ rectangular collection of dots, and rearrange the dots as indicated.

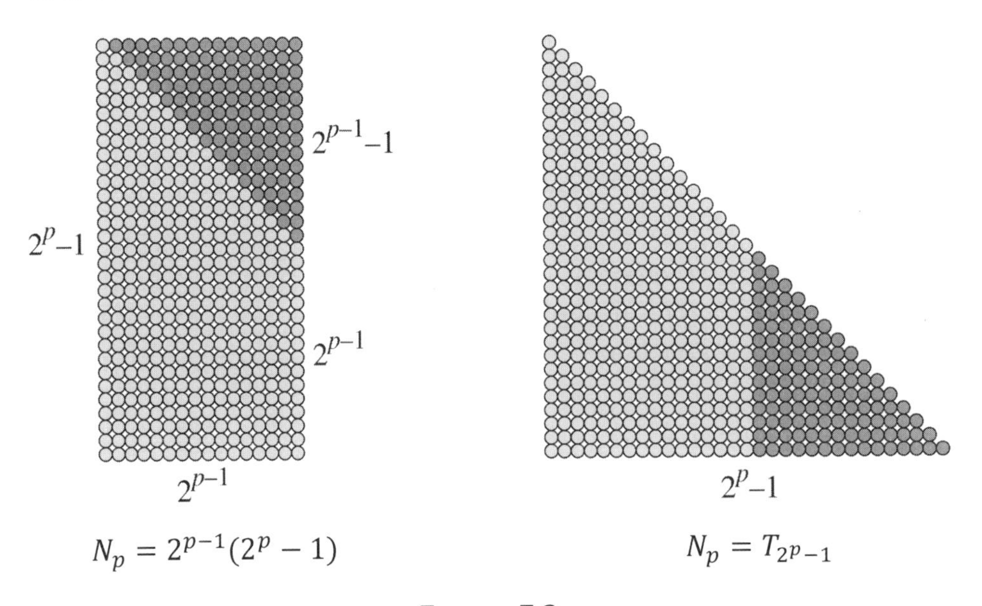

FIGURE 7.3

The algebraic proof is equally simple:

$$N_p = 2^{p-1}\left(2^p - 1\right) = \frac{2^p\left(2^p - 1\right)}{2} = T_{2^p-1}.$$

Hence we have

Theorem 7.3. *Every even perfect number is a triangular number.*

Even perfect numbers and Pascal's triangle modulo 2

Pascal's triangle modulo 2 is easily constructed using addition mod 2: $0+0=0$, $1+1=0$, and $1+0=1$. Here are the first 17 rows (the rows are numbered 0 to 16 counting from the top):

```
                1
               1 1
              1 0 1
             1 1 1 1
            1 0 0 0 1
           1 1 0 0 1 1
          1 0 1 0 1 0 1
         1 1 1 1 1 1 1 1
        1 0 0 0 0 0 0 0 1
       1 1 0 0 0 0 0 0 1 1
      1 0 1 0 0 0 0 0 1 0 1
     1 1 1 1 0 0 0 0 1 1 1 1
    1 0 0 0 1 0 0 0 1 0 0 0 1
   1 1 0 0 1 1 0 0 1 1 0 0 1 1
  1 0 1 0 1 0 1 0 1 0 1 0 1 0 1
 1 1 1 1 1 1 1 1 1 1 1 1 1 1 1 1
1 0 0 0 0 0 0 0 0 0 0 0 0 0 0 0 1
```

Central inverted triangular blocks of zeros (shaded gray) appear with the base in row $n=2^k$, and the number of zeros in the gray triangle is T_{2^k-1}, an even perfect number when k is prime. In the figure we see the perfect numbers $T_{2^2-1}=6$ and $T_{2^3-1}=28$, and the top row of T_{2^4-1} (which is not perfect). A row with exactly two 1's interpreted in binary (i.e., 11 equals 3, 101 equals 5, 10001 equals 17, etc.) are the Fermat numbers (see Exercise 2.8).

In Chapter 1 we showed that every hexagonal number H_n is a triangular number by illustrating $H_n = T_{2n-1} = n(2n-1)$. Hence each even perfect number is also a hexagonal number:

$$N_p = 2^{p-1}(2^p-1) = T_{2^p-1} = H_{2^{p-1}}.$$

In the next four sections we explore some of the many consequences of the representation of N_p as a triangular number. The results in Sections 7.4, 7.5, and 7.6 and Exercise 7.4 are motivated by the data in Table 7.1, where we evaluate N_p modulo 7, 9, and 12 for the first seven even perfect numbers.

TABLE 7.1

p	N_p	$N_p \pmod 7$	$N_p \pmod 9$	$N_p \pmod{12}$
2	6	-1	6	6
3	28	0	1	4
5	496	-1	1	4
7	8128	1	1	4
13	33550336	1	1	4
17	8589869056	-1	1	4
19	137438691328	1	1	4

7.4. Even perfect numbers modulo 9

Here is an interesting observation about even perfect numbers: If you sum the digits of any even perfect number (except 6), then sum the digits of the resulting number, and repeat this process until you get a single digit, that digit will be 1. For example, for the sixth even perfect number 8,589,869,056 we have

$$8,589,869,056 \to 64 \to 10 \to 1,$$

and for the seventh even perfect number 137,438,691,328 we have:

$$137,438,691,328 \to 55 \to 10 \to 1.$$

Summing digits of a number corresponds to evaluating the number modulo 9 since each power of 10 is congruent to 1 modulo 9. See Table 7.1 for some values of $N_p \pmod 9$, which leads to the following theorem.

Theorem 7.4. *Every even perfect number greater than 6 is congruent to* $1 \pmod 9$.

Proof. First note that when p is an odd prime, $2^p - 1 \equiv (-1)^p + 2 \equiv 1 \pmod 3$, so that $N_p = T_{2^p-1} = T_{3n+1}$ for some n. See Figure 7.4 for a visual proof that $T_{3n+1} \equiv 1 \pmod 9$. We also saw this result in Theorem 2.2. □

7.5. Even perfect numbers end in 6 or 28

Take another look at the seven even perfect numbers in Table 7.1—each one ends in 6 or 8, and those that end in 8 actually end in 28. We now show that this observation holds for every even perfect number. To do so, we prove the following lemma concerning T_{5n+1} and T_{5n+2}.

Lemma 7.5. $T_{5n+1} \equiv 1 \pmod 5$ *and* $T_{5n+2} \equiv 3 \pmod{25}$.

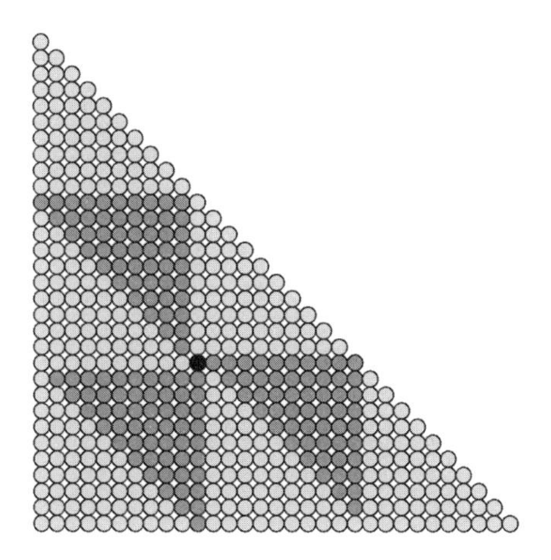

FIGURE 7.4. $N_p = T_{3n+1} = 1 + 9T_n$

Proof. In Figure 7.5(a) we show that $T_{5n+1} \equiv 1 \pmod 5$ by partitioning a triangular array of dots representing T_{5n+1} into 20 copies of T_n (in two shades of gray), five copies of T_{n-1} (white dots), and a single black dot. In Figure 7.5(b) we show that $T_{5n+2} \equiv 3 \pmod{25}$ by partitioning a triangular array of dots representing T_{5n+2} into 25 copies of T_n (in two shades of gray) and three black dots (we saw these results in Theorem 2.3). □

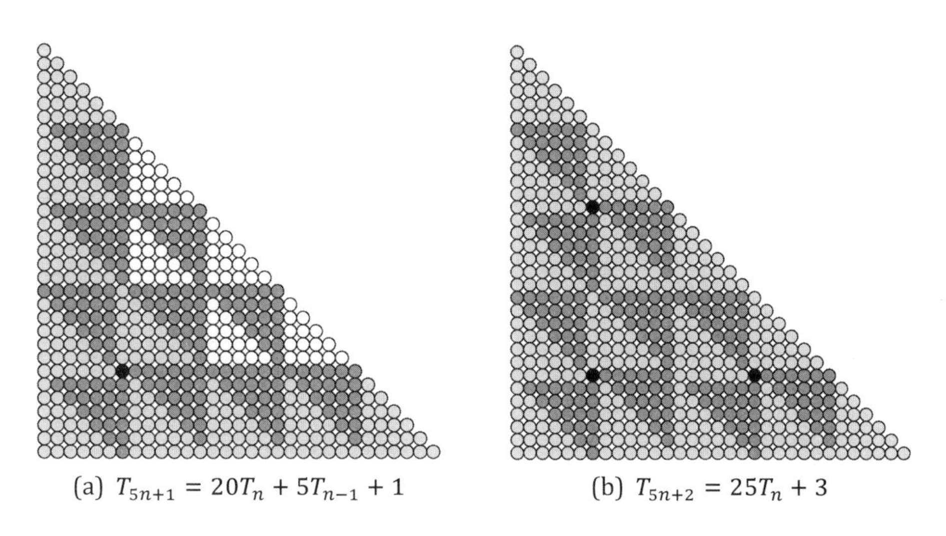

(a) $T_{5n+1} = 20T_n + 5T_{n-1} + 1$ (b) $T_{5n+2} = 25T_n + 3$

FIGURE 7.5

Theorem 7.6. *Every even perfect number ends in* 6 *or* 28.

Proof. Since $N_2 = 6$ we need only consider N_p for p odd. When p is odd there are two cases: $p = 4k + 1$ and $p = 4k + 3$. If $p = 4k + 1$, then

$2^p - 1 = 2 \cdot 16^k - 1 \equiv 1 \pmod 5$, i.e., $2^p - 1 = 5n + 1$ for some positive integer n. Hence $N_p = T_{2^p-1} = T_{5n+1} \equiv 1 \pmod 5$, so that in base 10, N_p ends in 1 or 6. Since N_p is even, it ends in 6.

If $p = 4k + 3$, then $2^p - 1 = 8 \cdot 16^k - 1 \equiv 2 \pmod 5$, i.e., $2^p - 1 = 5n + 2$ for some positive integer n. Hence $N_p = T_{2^p-1} = T_{5n+2} \equiv 3 \pmod{25}$, so that in base 10, N_p ends in 03, 28, 53, or 78. Since N_p is a multiple of 4 for $p \geq 3$, it ends in 28. □

7.6. Even perfect numbers modulo 7

With the lone exception of $N_3 = 28$, every even perfect number is congruent to 1 or −1 modulo 7 (see Table 7.1). To prove this result [Wall, 1984] , we use the fact that every prime $p \neq 3$ is congruent to 1 or 2 modulo 3.

Theorem 7.7. *Every even perfect number N_p with $p \neq 3$ is congruent to 1 or −1 modulo 7, i.e.,*

$$p \equiv 1 \pmod 3 \Rightarrow N_p \equiv 1 \pmod 7, \text{ and}$$
$$p \equiv 2 \pmod 3 \Rightarrow N_p \equiv -1 \pmod 7.$$

Proof. If $p = 3k + 1$, then $2^p - 1 = 2 \cdot 8^k - 1 \equiv 1 \pmod 7$, so that $N_{3k+1} = T_{7n+1}$. In Figure 7.6(a) we show that $T_{7n+1} \equiv 1 \pmod 7$ by partitioning a triangular array of dots representing T_{7n+1} into 35 copies of T_n (in two shades of gray), 14 copies of T_{n-1} (white dots), and a single black dot. Thus $N_{3k+1} \equiv 1 \pmod 7$.

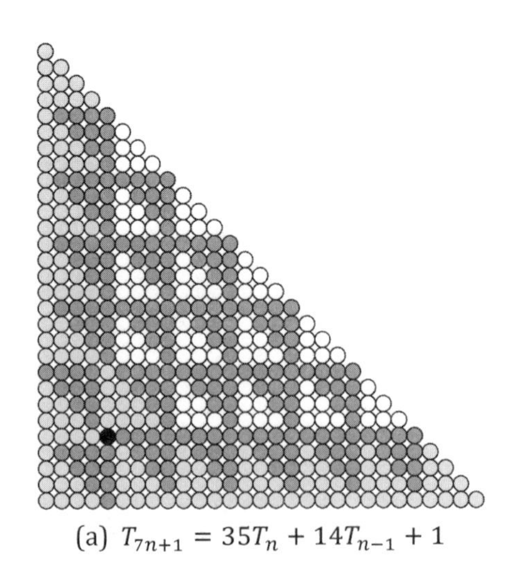

(a) $T_{7n+1} = 35T_n + 14T_{n-1} + 1$

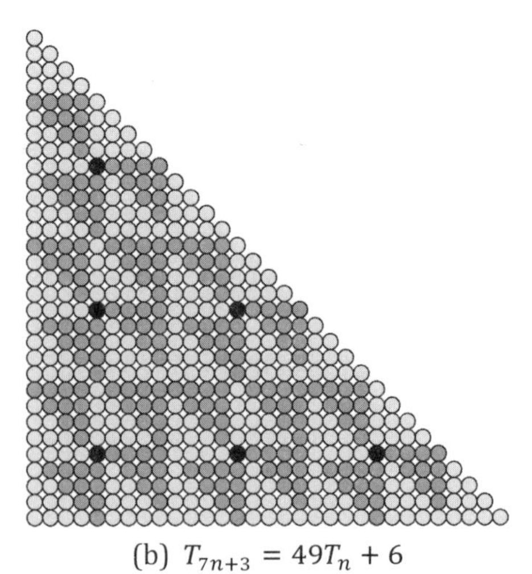

(b) $T_{7n+3} = 49T_n + 6$

FIGURE 7.6

If $p = 3k+2$, then $2^p - 1 = 4 \cdot 8^k - 1 \equiv 3 \pmod 7$, so that $N_{3k+2} = T_{7n+3}$. In Figure 7.6(b) we see that $T_{7n+3} \equiv 6 \pmod 7$ by partitioning a triangular array of dots representing T_{7n+3} into 49 copies of T_n (in two shades of gray) and six black dots. Hence $N_{3k+2} \equiv 6 \pmod 7$ or, equivalently, $N_{3k+2} \equiv -1 \pmod 7$. □

7.7. Even perfect numbers and sums of odd cubes

Observing that $N_3 = 28 = 1^3 + 3^3$, $N_5 = 496 = 1^3 + 3^3 + 5^3 + 7^3$, and $N_7 = 8128 = 1^3 + 3^3 + 5^3 + \cdots + 15^3$ yields the hypothesis for the following theorem.

Theorem 7.8. *Every even perfect number N_p with $p \geq 3$ is the sum of the first n odd cubes for $n = 2^{(p-1)/2}$, i.e., $N_p = 1^3 + 3^3 + \cdots + (2n-1)^3$. Note that the last integer cubed is $2^{(p+1)/2} - 1$.*

Proof. When $n = 2^{(p-1)/2}$ we have $2n^2 = 2^p$, so that $N_p = T_{2^p-1} = T_{2n^2-1}$. In Figure 7.7 we show that $T_{2n^2-1} = 1 \cdot 1^2 + 3 \cdot 3^2 + \cdots + (2n-1) \cdot (2n-1)^2$ (for $p = 5$ and $n = 4$), representing k^3 as k copies of k^2 dots. The formula for T_{2n^2-1} as a sum of odd cubes can also be proved by mathematical induction. Hence N_p (for p odd) is the sum of the first $n = 2^{(p-1)/2}$ odd cubes. □

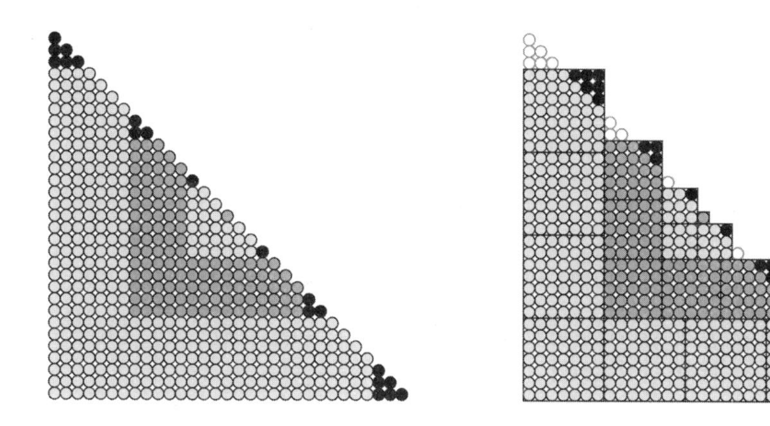

FIGURE 7.7

7.8. Odd perfect numbers

As noted at the beginning of this chapter, whether or not odd perfect numbers exist is an open problem. Nonetheless, in this section we prove the following simple result about odd perfect numbers: *If N is an odd perfect number, then N has at least three distinct prime factors.*

It is common in number theory to let $\sigma(n)$ denote the sum of all the divisors of n, including n itself. Hence n is perfect if only if $\sigma(n) = 2n$ or, equivalently, $\sigma(n)/n = 2$.

To show that an odd integer n with just one or two distinct prime factors cannot be perfect, we use the following lemma.

Lemma 7.9. *If r is a real number such that $0 < r < 1$ and k is any positive integer, then*

$$1 + r + r^2 + \cdots + r^k < \frac{1}{1-r}.$$

Proof. See Figure 7.8, and observe that the area of the darker gray rectangle is less than the area of the entire rectangle in two shades of gray. □

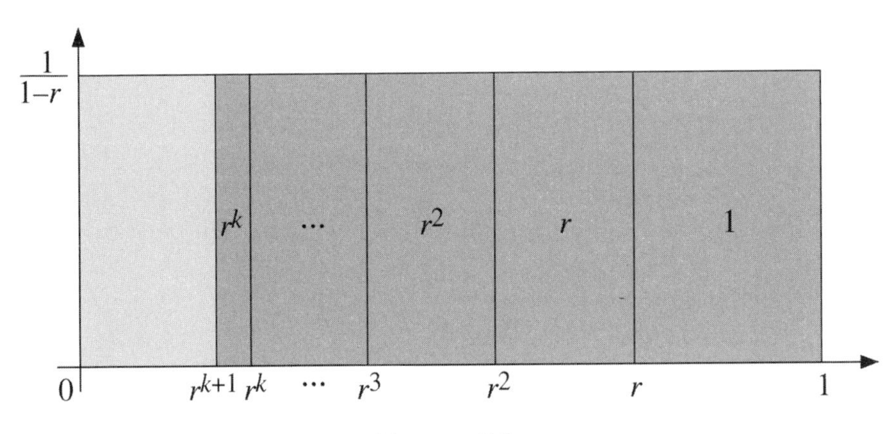

FIGURE 7.8

Theorem 7.10. *If $N = p^n$ or $N = p^n q^m$, where p and q are distinct odd primes and m and n are positive integers, then N is not perfect.*

Proof. Let $N = p^n$. Then the divisors of n are $1, p, p^2, \ldots, p^n$, and since $0 < 1/p < 1$ with $p \geq 3$ we have

$$\begin{aligned}\frac{\sigma(p^n)}{p^n} &= \frac{1}{p^n}\left(1 + p + p^2 + \cdots + p^n\right) \\ &= 1 + \frac{1}{p} + \left(\frac{1}{p}\right)^2 + \cdots + \left(\frac{1}{p}\right)^n \\ &< \frac{1}{1-(1/p)} \leq \frac{1}{1-(1/3)} = \frac{3}{2} < 2,\end{aligned}$$

so that $N = p^n$ is not perfect. The σ function, like many number theoretic functions, is *multiplicative*, that is, if m and n are relatively prime positive integers, then $\sigma(mn) = \sigma(m)\sigma(n)$. Now let $N = p^n q^m$. If p and q are distinct odd primes, then one of them (say p) is at least 3, and the other (q) is at least 5. Hence

$$\frac{\sigma(q^m)}{q^m} < \frac{1}{1-(1/q)} \leq \frac{1}{1-(1/5)} = \frac{5}{4};$$

consequently

$$\frac{\sigma(p^n q^m)}{p^n q^m} = \frac{\sigma(p^n)}{p^n} \cdot \frac{\sigma(q^m)}{q^m} < \frac{3}{2} \cdot \frac{5}{4} = \frac{15}{8} < 2,$$

and thus $N = p^n q^m$ is not perfect. □

7.9. Exercises

7.1 Show that every even perfect number greater than six is the sum of three distinct triangular numbers, e.g., $N_5 = 496 = 231 + 210 + 55 = T_{21} + T_{20} + T_{10}$.

7.2 Here is another pattern relating even perfect numbers and triangular numbers:

$$N_2 = 6 = 2 \cdot 3 = 3 \cdot 1 + 3 = 3T_1 + T_2,$$
$$N_3 = 28 = 4 \cdot 7 = 3 \cdot 6 + 10 = 3T_3 + T_4,$$
$$N_5 = 496 = 16 \cdot 31 = 3 \cdot 120 + 136 = 3T_{15} + T_{16},$$
$$N_7 = 8128 = 64 \cdot 127 = 3 \cdot 2016 + 2080 = 3T_{63} + T_{64}, \text{ etc.}$$

State and prove a theorem about even perfect numbers.

7.3 Here is a third pattern:

$$N_3 = 28 = 1 \cdot 10 + 3 \cdot 6 = T_1 T_4 + T_2 T_3,$$
$$N_5 = 496 = 6 \cdot 36 + 10 \cdot 28 = T_3 T_8 + T_4 T_7,$$
$$N_7 = 8128 = 28 \cdot 136 + 36 \cdot 120 = T_7 T_{16} + T_8 T_{15}, \text{ etc.}$$

State and prove a theorem about even perfect numbers greater than 6.

7.4 If p is an odd prime, show that $N_p \equiv 4 \pmod{12}$. [Hint: See Theorem 7.4.]

7.5 As a consequence of the preceding exercise, in base 12 every even perfect number greater than 6 ends in 4. Show that more is true—in base 12 every even perfect number greater than 28 ends in 54.

7.6 Express N_p in binary (base 2) notation.

7.7 Show that if N_p is an even perfect number, then $\sqrt{1 + 8N_p}$ is an integer, e.g., $\sqrt{1 + 8 \cdot 6} = 7$, $\sqrt{1 + 8 \cdot 28} = 15$, $\sqrt{1 + 8 \cdot 496} = 63$, etc.

7.8 Prove that if N_p ends in 6 for $p \geq 3$, then the digit preceding 6 must be odd.

7.9 Even perfect numbers are constructed from the *Mersenne primes*, primes of the form $M_p = 2^p - 1$ where p is also prime. The first few are $3, 7, 31, 127, 8191, \ldots$.

(a) Show that for $p \geq 3$, $M_p \equiv 1 \pmod 6$ and $M_p \equiv 7 \pmod{24}$.

(b) Show that for $p \geq 3$, the units digit (in base 10) of M_p is 1 or 7.

7.10 Can a Mersenne prime be the hypotenuse element of a PPT?

Solutions to the Exercises

Chapter 1

1.1 See Figure S1.1 for illustrations of (a) $4^2+O_4 = T_8$ and (b) $5^2+O_4 = T_9$.

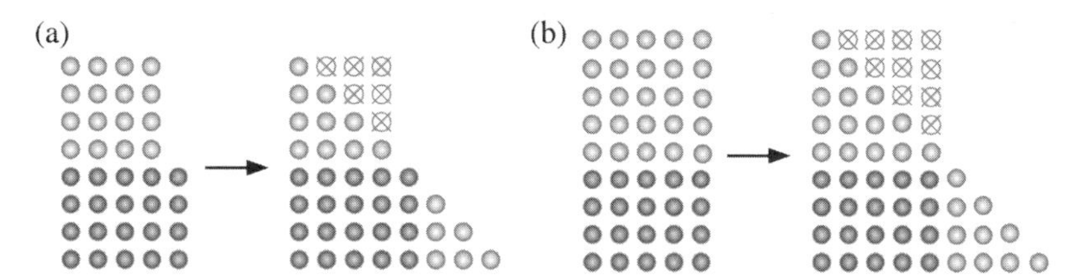

FIGURE S1.1

1.2 See Figure S1.2, where we illustrate $1 + 9 + 9^2 = T_{1+3+3^2}$.

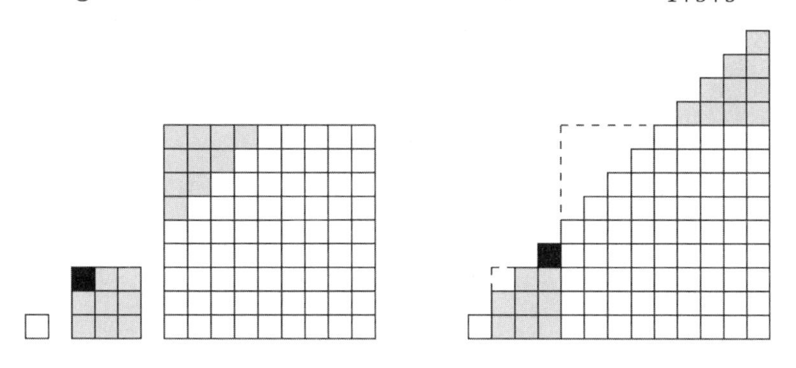

FIGURE S1.2

1.3 Yes, there are infinitely many more, since (1.4) is equivalent to $T_n = 2T_p \Leftrightarrow T_{3n+4p+3} = 2T_{2n+3p+2}$.

1.4 In Figure S1.3 we illustrate $(2n+1)^2 = T_{3n+1} - T_n$ for $n = 3$.

1.5 The kth n-gonal number is

$$\frac{k}{2}\left[(n-2)\,k-(n-4)\right] = \frac{k}{2}\left[(n-2)\,(k-1)+2\right].$$

Assume $k \geq 3$ and $n \geq 3$. When k is even the number factors as $\frac{k}{2} \cdot [(n-2)\,(k-1)+2]$ with both factors integers greater than 1.

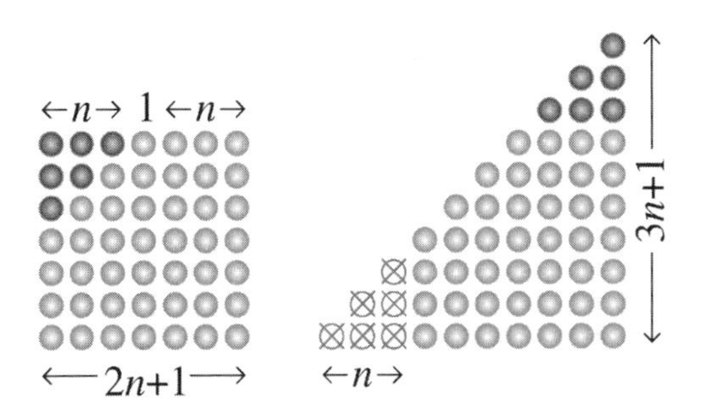

FIGURE S1.3

When k is odd, $k-1$ is even and the number factors as $k \cdot \frac{(n-2)(k-1)+2}{2}$ with both factors integers greater than 1.

1.6 See Figure S1.4.

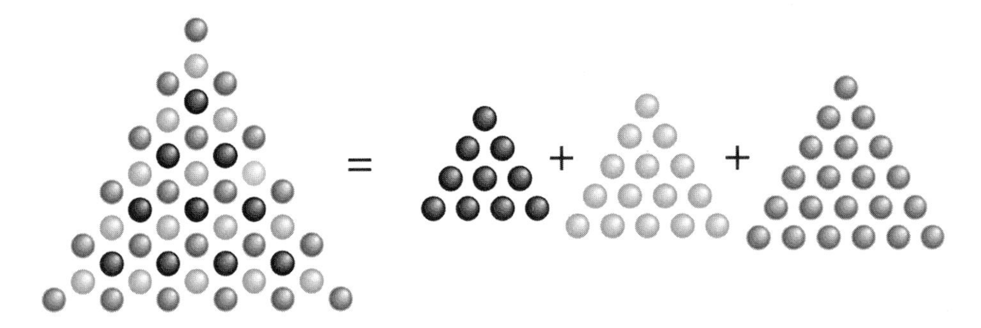

FIGURE S1.4

1.7 Using the hint, we have $1^2+3^2+5^2+\cdots+(2n-1)^2 = T_1 + (T_2+T_3)+(T_4+T_5)+\cdots+(T_{2n-2}+T_{2n-1}) = \mathrm{Tet}_{2n-1}$.

1.8 The theorem is: *Four times the nth pyramidal number is the* $(2n)$*th tetrahedral number*; i.e., $4\,\mathrm{Pyr}_n = \mathrm{Tet}_{2n}$, and a proof is

$$4\frac{n(n+1)(2n+1)}{6} = \frac{(2n)(2n+1)(2n+2)}{6}.$$

You can illustrate this result by restacking the tetrahedral pile of cannonballs into four pyramidal piles, as indicated here for the eighth tetrahedral number 120:

$$\begin{aligned}120 &= (1+3)+(6+10)+(15+21)+(28+36)\\ &= 4+16+36+64 = 4(1+4+9+16).\end{aligned}$$

Hence Pyr_n is $\frac{1}{4}$ of a binomial coefficient:

$$\mathrm{Pyr}_n = \frac{1}{4}\mathrm{Tet}_{2n} = \frac{1}{4}\binom{2n+2}{3}.$$

1.9 The nth octahedral number is the sum of the $(n-1)$st and nth pyramidal numbers and hence it equals

$$\frac{(n-1)n(2n-1)}{6}+\frac{n(n+1)(2n+1)}{6}=\frac{n(2n^2+1)}{3}.$$

Chapter 2

2.1 Placing a 1 at the end of a number k in base 9 produces the number $9k+1$. If $k=T_n$, then $9T_n+1=T_{3n+1}$ from Theorem 2.2. Since 1 is triangular, so are 11_9, 111_9, etc. [Also see Exercise 1.2 and its solution.]

2.2 The $(2k+1)$-st n-gonal number is $(2k+1)+(n-2)T_{2k}$, but $T_{2k}=k(2k+1)$, so the $(2k+1)$-st n-gonal number is a multiple of $2k+1$. Similarly, the $(2k)$th n-gonal number is a multiple of k (but not necessarily $2k$).

2.3 In Theorem 1.3 for
   (a) set $(n,a,b,c)=(3k-1,2k,2k,2k)$,
   (b) set $(n,a,b,c)=(3k,2k,2k,2k)$, and
   (c) set $(n,a,b,c)=(3k+1,2k+1,2k+1,2k+1)$.

2.4 From the proof of Theorem 2.2 we have $T_{3n+1}=9T_n+1$ and $T_{3n-1}\equiv T_{3n}\equiv 0 \pmod 3$. Hence for every positive integer t, t is triangular if and only if $9t+1$ is triangular. This yields the sequence $(a,b)=(9,1),(81,10),(729,91),\ldots$, i.e., $(a,b)=(9^k,(9^k-1)/8)$ for $k=1,2,3,\ldots$.

2.5 Observe that n^p-n is even and p and 2 are relatively prime, hence n^p-n is a multiple of $2p$.

2.6 We need to show that $n^3\equiv n \pmod 6$. Set $p=3$ in the preceding exercise.

2.7 We need to show that $n^5\equiv n \pmod{10}$. Set $p=5$ in Exercise 2.5.

2.8 It suffices to show that 2^{2^n} ends in 6 for $n\geq 2$ and that its tens digit is odd. Since the exponent 2^n is a multiple of 4 when $n\geq 2$, we need only show that 2^{4k} ends in 6 for $k\geq 1$. Setting $p=5$ and $n=2^k$ in Corollary 2.10 yields $2^{4k}=(2^k)^4\equiv 1 \pmod 5$. So (a) 2^{4k} ends in 1 or 6, but since it is even, it ends in 6, and (b) since 2^{4k} is divisible by 4, it ends with 16, 36, 56, 76, or 96, i.e., the tens digit is odd.

2.9 It suffices to show that $2^{2^n}\equiv 4 \pmod{12}$ for $n\geq 1$. In this case the exponent 2^n is even, so $2^{2^n}\equiv(-1)^{2^n}\equiv 1 \pmod 3$, so $2^{2^n}\equiv 1$, 4, 7, or 10 $\pmod{12}$. But 2^{2^n} is a multiple of 4, hence $2^{2^n}\equiv 4 \pmod{12}$.

2.10 Yes, since the sum in question is the nth tetrahedral number from Section 1.5:

$$\mathrm{Tet}_n = \sum_{k=1}^{n} T_k = \frac{n(n+1)(n+2)}{6} = T_n \cdot \frac{n+2}{3}.$$

Hence $\sum_{k=1}^{n} T_k \equiv 0 \pmod{T_n}$ if and only if $\frac{n+2}{3}$ is an integer, i.e., $n \equiv 1 \pmod 3$.

2.11 Since $P_n = T_{3n-1}/3$, we require $T_{3n-1} \equiv 0 \pmod 5$. From Theorem 2.3, $3n - 1 \equiv 0$ or $4 \pmod 5$, hence $3n \equiv 0$ or $1 \pmod 5$ so that $n \equiv 0$ or $2 \pmod 5$.

2.12 Each term in the expansion of the given determinant is, except for sign, the product of all possible row indices and all possible column indices, that is, $(100!)^2$, and this is the absolute value of every term. Now 101 is a prime, so by Wilson's theorem, $100! \equiv -1 \pmod{101}$. Hence $(100!)^2 \equiv (-1)^2 \equiv 1 \pmod{101}$, as required.

2.13 Suppose a prime $q < p$ divides $(p-1)!+1$. Since q divides $(p-1)!$ it also divides $(p-1)!+1-(p-1)! = 1$, which is impossible. But p divides $(p-1)!+1$ by Wilson's theorem, hence it is the smallest prime that does so.

Chapter 3

3.1 With five sailors we have $x = 5a + 1$, $4a = 5b + 1$, $4b = 5c + 1$, $4c = 5d + 1$, $4d = 5e + 1$, and $4e = 5y + 1$. Eliminating a, b, c, d, and e yields the Diophantine equation $1024x - 15625y = 11529$. Since $(x, y) = (-4776, -313)$ is a solution to $1024x - 15625y = 1$, $(x_0, y_0) = (-4776 \cdot 11529, -313 \cdot 11529)$ is a solution to $1024x - 15625y = 11529$. Hence the general solution is $(x, y) = (-4776 \cdot 11529 + 15625k, -313 \cdot 11529 + 1024k)$. For $k \geq 3525$ we have positive solutions, the smallest of which is $(x, y) = (15621, 1023)$.

3.2 Yes. Let $\{p_k\}_{k=1}^{1{,}000{,}000}$ be a set of 1,000,000 distinct primes. Then p_i^2 and p_j^2 are relatively prime when $i \neq j$, so by the Chinese remainder theorem, there is an integer x such that $x \equiv -k \pmod{p_k^2}$ for $1 \leq k \leq 1{,}000{,}000$. So $x + k$ is divisible by p_k^2 (i.e., $x + k$ has a repeated prime factor). Thus $\{x + k\}_{k=1}^{1{,}000{,}000}$ is a set of 1,000,000 consecutive integers each of which contains a repeated prime factor.

3.3 Use mathematical induction: $x_1 y_1 = 6$ and $x_{n+1} y_{n+1} = 6x_n^2 + 17x_n y_n + 12y_n^2$.

3.4 If $x_n^2 - 2y_n^2 = +1$, then $x_n^2 y_n^2 = \left(2y_n^2 + 1\right) y_n^2 = T_{2y_n^2}$. If $x_n^2 - 2y_n^2 = -1$, then $x_n^2 y_n^2 = x_n^2 \left[\left(x_n^2 + 1\right)/2\right] = T_{x_n^2}$.

3.5 The problem asks for integers k and m such that $1 + 2 + \cdots + (k-1) = (k+1) + (k+2) + \cdots + m$. This simplifies to $T_{k-1} = T_m - T_k$, or $T_m = T_{k-1} + T_k$ so that $T_m = k^2$. Since $50 \le m \le 500$, we see that $k = 204$ from Example 3.6 and Table 3.3.

3.6 *Proof* 1. Solutions (x_n, y_n) to the Pell equation $x^2 - 2y^2 = 1$ in Table 3.3 yield the pairs $(2y_n^2, x_n^2)$ of consecutive powerful numbers, e.g., the solution (99,70) for $n = 3$ yields the pair (9800,9801). Similarly, solutions (x_n, y_n) to the Pell equation $x^2 - 2y^2 = -1$ in Table 3.4 yield the pairs $\left(x_n^2, 2y_n^2\right)$ of consecutive powerful numbers, e.g., the solution (41,29) for $n = 2$ yields the pair (1681,1682).

Proof 2. Let $T_n = k^2$ be a square triangular number, and consider the pair $(8T_n, 8T_n + 1)$. In terms of n and k, this is the pair $(8k^2, (2n+1)^2)$ using Lemma 2.4. If a prime p divides $8k^2$ or $(2n+1)^2$, so does p^2, and the result follows since there are infinitely many square triangular numbers. However, not all pairs of consecutive powerful numbers are obtained this way, e.g., $(675, 676) = (3^3 \cdot 5^2, 2^2 \cdot 13^2)$.

3.7 Let (x_n, y_n) be one of the infinitely many solutions in integers to the Pell equation $x^2 - 2y^2 = 1$ in Table 3.3. Then $x_n^2 - 1$ $(= y_n^2 + y_n^2)$, $x_n^2 + 0$, $x_n^2 + 1$ satisfy the requirements of the problem.

3.8 Yes, infinitely many. The equation $(y-1)^2 + y^2 + (y+1)^2 = x^2 + 1$ simplifies to $x^2 - 3y^2 = 1$. See Section 3.5.

3.9 Yes, infinitely many. The equation $T_m = 3T_k$ is equivalent to the Pell equation $x^2 - 3y^2 = -2$ with $x = 2m + 1$ and $y = 2k + 1$. Hence the pairs (x_n, y_n) in Table 3.6 yield solutions, e.g., $(x_3, y_3) = (71, 41)$ yields $(m, k) = (35, 20)$, $(x_4, y_4) = (265, 153)$ yields $(m, k) = (132, 76)$, etc.

3.10 The kth octagonal number Oct_k is $k^2 + 4T_{k-1} = k(3k-2)$, and the equation $k(3k-2) = m^2$ yields the Pell equation $(3k-1)^2 - 3m^2 = 1$. So solutions (x_n, y_n) in Table 3.5 with $x_n \equiv -1 \pmod 3$ yield square octagonal numbers, e.g., $(x_3, y_3) = (26, 15)$ yields $\mathrm{Oct}_9 = 15^2$, $(x_5, y_5) = (362, 209)$ yields $\mathrm{Oct}_{121} = 209^2$, etc.

3.11 Yes, infinitely many. The equation $S_m = k^2$ is equivalent to the Pell equation $(2k)^2 - 6(2m-1)^2 = -2$. The seed $(x_0, y_0) = (2, 1)$ with $x_n^2 - 6y_n^2 = -2$ generates infinitely many solutions, e.g., $(x_1, y_1) = (22, 9)$ yields $S_5 = 11^2$, $(x_2, y_2) = (218, 89)$ yields $S_{45} = 109^2$, $(x_3, y_3) = (2158, 881)$ yields $S_{441} = 1079^2$, etc.

3.12 $T_m = k^2 - 1$ is equivalent to $(2m+1)^2 - 2(2k)^2 = -7$. The seed $(x_0, y_0) = (1, 2)$ with $x_n^2 - 2y_n^2 = -7$ generates infinitely many solutions which lead to $T_5 = 4^2 - 1$, $T_{32} = 23^2 - 1$, etc.; while the seed $(x_0, y_0) = (-1, 2)$ leads to $T_2 = 2^2 - 1$, $T_{15} = 11^2 - 1$, etc., $T_m = k^2 + 1$ is equivalent to $(2m+1)^2 - 2(2k)^2 = 9$. The seed $(x_0, y_0) = (3, 0)$ with $x_n^2 - 2y_n^2 = 9$ generates infinitely many solutions which lead to $T_4 = 3^2+1$, $T_{25} = 18^2+1$, $T_{148} = 105^2+1$, etc.

3.13 Yes. Consider the equation $m^2 = (k+1)^3 - k^3 = 3k^2 + 3k + 1$. Completing the square yields $(2m)^2 - 3(2k+1)^2 = 1$. So pairs (x_n, y_n) in Table 3.5 with x_n even and y_n odd yield solutions, e.g., $(x_3, y_3) = (26, 15)$ yields $(m, k) = (13, 7)$, $(x_5, y_5) = (362, 209)$ yields $(m, k) = (181, 104)$ so that $181^2 = 105^3 - 104^3$, etc.

3.14 If $y^2 = k + 1$ and $x^2 = 3k + 1$, then $x^2 - 3y^2 = -2$, and solutions (x_n, y_n) to this Pell equation appear in Table 3.6. Then $k = (x_n^2 - y_n^2)/2$ leads to the solutions $k = 8, 120, 1680, \ldots$.

3.15 Yes. $T_{m-1} + T_m + T_{m+1} = k^2$ is equivalent to $(4k)^2 - 6(2m+1)^2 = 10$. Solving the Pell equation $x^2 - 6y^2 = 10$ with initial solution $(x_0, y_0) = (4, 1)$ yields solutions $T_{63} + T_{64} + T_{65} = 79^2$, $T_{637} + T_{638} + T_{639} = 782^2$, etc.

3.16 See Figure S3.1 for a visual proof that $K = rs$, using the fact that the three angle bisectors in a triangle meet at the center of the inscribed circle.

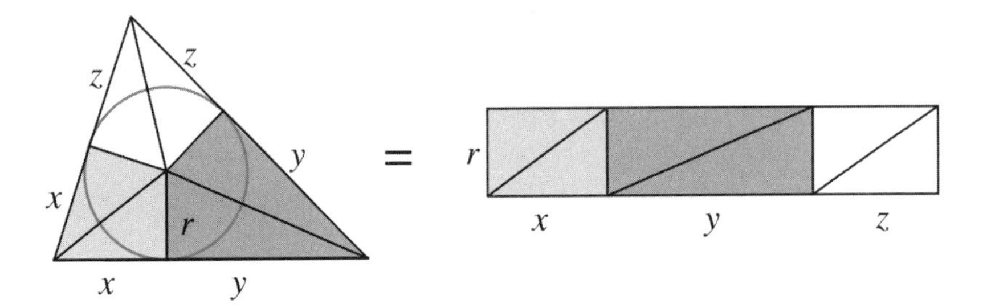

FIGURE S3.1

For an almost equilateral Heronian triangle in Example 3.5, $K = 3x_n y_n$ and $s = 3x_n$, hence $r = y_n$, an integer.

3.17 The probability that both balls are red is $1/2 = \binom{r}{2}/\binom{k}{2}$, so that $2r(r-1) = k(k-1)$. Completing the square yields $(2k-1)^2 - 2(2r-1)^2 = -1$, the Pell equation $x^2 - 2y^2 = -1$ with $x = 2k - 1$ and $y = 2r - 1$. Each solution (x_n, y_n) for $n \geq 1$ in Table 3.4 has x and y odd and $y > 2$, and hence leads to a solution. For example,

$(x_1, y_1) = (7, 5)$ leads to $(k, r) = (4, 3)$, $(x_2, y_2) = (41, 29)$ leads to $(k, r) = (21, 15)$, $(x_3, y_3) = (239, 169)$ leads to $(k, r) = (120, 85)$, etc.

Chapter 4

4.1 If a and b are both even, then so is c, so (a, b, c) cannot be primitive. If a and b are both odd, then $a^2 \equiv 1 \pmod 4$, $b^2 \equiv 1 \pmod 4$, and so $c^2 \equiv 2 \pmod 4$, which is impossible since even squares are congruent to $0 \pmod 4$. Hence a and b have opposite parity, and c is odd.

4.2 $2mn$ is a multiple of 4 since either m or n is even. Now consider the generator $(m, n) = (2k, 1)$ for $k \geq 1$.

4.3 When $c = m^2 + n^2$, $2c = 2\left(m^2 + n^2\right) = (m+n)^2 + (m-n)^2$, and both $m + n$ and $m - n$ are odd since m and n have opposite parity.

4.4 Yes, infinitely many. The generator $(m, n) = (T_k, T_{k-1})$ for $k \geq 2$ generates the PT $(k^3, T_{k^2-1}, T_{k^2})$. None are primitive, since each side is a multiple of k.

4.5 The generator $(m, n) = \left(2^{2^k}, 1\right)$ for $k \geq 0$ generates the PPT $(f_{k+1} - 2, 2(f_k - 1), f_{k+1})$.

4.6 $T_m = k(k+1)$ implies $m^2 + (m+1)^2 = 2m^2 + 2m + 1 = 4k^2 + 4k + 1 = (2k+1)^2$, and conversely $m^2 + (m+1)^2 = (2k+1)^2$ implies $m(m+1) = 2k(k+1)$ so that $T_m = k(k+1)$.

4.7 The generators $(m, n) = (p+1, 1)$ and $(m, n) = (p+1, p)$ generate different PPTs with inradius p.

4.8 The generator $(m, 1)$ generates a PPT with area

$$(m-1)\,m(m+1).$$

4.9 Yes, infinitely many. The PPT $\left(2k+1, 2k^2+2k, 2k^2+2k+1\right)$ from Example 4.1 has $P = 4k^2 + 6k + 2 = 2T_{2k+1}$ and

$$A = k\,(k+1)\,(2k+1) = 6\mathrm{Pyr}_k.$$

4.10 Express the inradius and exradii in terms of the generator (m, n) and simplify.

4.11 Using the hint we have

$$\begin{aligned}(ab)^4 + (bc)^4 + (ac)^4 &= a^4b^4 + c^4\left(a^4 + b^4\right)\\ &= a^4b^4 + c^4\left(a^2 + b^2\right)^2 - 2c^4a^2b^2\\ &= c^8 - 2c^4a^2b^2 + a^4b^4\\ &= (c^4 - a^2b^2)^2.\end{aligned}$$

4.12 Setting $q = 0$ in Theorem 4.19 yields

$$(m^2 + n^2 + p^2)^2 = (m^2 + n^2 - p^2)^2 + (2mp)^2 + (2np)^2.$$

Note: Adrien-Marie Legendre (1752–1833) proved that an integer is the sum of three positive integer squares if and only if it is not of the form $4^n(8k + 7)$ for non-negative integers k and n.

4.13 There are infinitely many such sets of three squares. The sides of the three squares in Figure 4.24 are $b - a$, c, and $a + b$, and hence their areas $(b - a)^2 = c^2 - 2ab$, c^2, and $(a + b)^2 = c^2 + 2ab$ form an arithmetic progression of three squares with a common difference equal to four times the area of the PΔ. The two examples in the statement of the problem are generated by the PΔ's (3,4,5) and (7,24,25).

4.14 Yes, infinitely many. From Lemma 4.21 we have

$$(a^2 + b^2)^2 = (a^2 - b^2)^2 + (2ab)^2,$$

so that

$$\begin{aligned}(a^2 + b^2)^3 &= \left[(a^2 - b^2)^2 + (2ab)^2\right](a^2 + b^2)\\ &= (a^3 + ab^2)^2 + (a^2b + b^3)^2\\ &= (a^3 - 3ab^2)^2 + (3a^2b - b^3)^2.\end{aligned}$$

Chapter 5

5.1 Using the hint we have $a^2 + b^2 \equiv 0 \pmod 3$, hence $a \equiv b \equiv 0 \pmod 3$ from Example 3.1. This contradicts a and b relatively prime.

5.2 (a) The Gelfand-Schneider number $2^{\sqrt{2}}$ is an example.
(b) $1^{\sqrt{2}} = 1$ is an example. As a consequence of the solution to Hilbert's seventh problem, we cannot provide an example showing that a rational not equal to 0 or 1 to an irrational power may be rational.

5.3 Assume $\sqrt{n^2 - 1}$ is irrational, and write $\sqrt{n^2 - 1} = a/b$ in lowest terms for positive integers a and b. Then $a^2 + b^2 = (nb)^2$. Following the hint we modify Figure 5.6 so that the large right triangle has hypotenuse nb, vertical leg b, and horizontal leg a. Consequently, the smaller gray triangle has legs $nb - a$ and $na - (n^2 - 1)b$, and hypotenuse $n^2b - na$. Thus $\sqrt{n^2 - 1} = [na - (n^2 - 1)b]/(nb - a)$, a contradiction since $na - (n^2 - 1)b$ and $nb - a$ are smaller than a and b, respectively. Hence $\sqrt{n^2 - 1}$ is irrational.

5.4 Assume $\sqrt{n^2+1}$ is irrational, and write $\sqrt{n^2+1} = a/b$ in lowest terms for positive integers a and b. Then $b^2 + (nb)^2 = a^2$. Following the hint we modify Figure 5.8 so that the large right triangle has hypotenuse a, vertical leg b, and horizontal leg nb. Consequently the smaller gray triangle has legs $a - nb$ and $na - n^2b$, and hypotenuse $(n^2+1)b - na$. Thus $\sqrt{n^2-1} = [(n^2+1)b - na]/(a - nb)$, a contradiction since $(n^2+1)b - na$ and $a - nb$ are smaller than a and b, respectively. Hence $\sqrt{n^2+1}$ is irrational.

5.5 (a) Create a square with area 3 in Figure 5.12 by drawing a diagonal in each of the four rectangles.
(b) Partition the square with side length $\varphi+1+\varphi$ into a unit square surrounded by four $\varphi \times (\varphi+1)$ rectangles. Hence $(2\varphi+1)^2 = 1 + 4(\varphi^2 + \varphi)$. But $\varphi^3 = \varphi^2 + \varphi = 2\varphi + 1$ so that $(\varphi^3)^2 = 1 + 4\varphi^3$ from which the result follows.

5.6 (a) Since $\varphi_n = \frac{1}{2}(n + \sqrt{n^2+4})$, $n\varphi_n = \frac{1}{2}\left(n^2 + \sqrt{n^4+4n^2}\right) = \left[\overline{n^2, 1}\right]$. Thus $\varphi_n^2 = 1 + n\varphi_n = 1 + \left[\overline{n^2, 1}\right] = \left[1 + n^2, \overline{1, n^2}\right]$.
(b) Expanding and simplifying $\left(\varphi_n - \frac{1}{\varphi_n}\right)^3 = n^3$ yields $\varphi_n^3 - \frac{1}{\varphi_n^3} = n^3 + 3n$, hence $\varphi_n^3 = \varphi_{n^3+3n} = \left[\overline{n^3+3n}\right]$.

5.7 Yes. It is

$$\sin\frac{\pi}{2^{n+1}} = \frac{1}{2}\sqrt{2 - \sqrt{2 + \sqrt{2 + \cdots + \sqrt{2}}}}$$

with n square root signs. This follows from $\sin\theta = \sqrt{1 - \cos^2\theta}$ with $\theta = \frac{\pi}{2^{n+1}}$.

5.8 Using the hint we have $x = \sqrt{a + \sqrt{a + \sqrt{a + \cdots}}} = \sqrt{a + x}$, so $x^2 - x - a = 0$. Hence $x = \frac{1}{2}[1 + \sqrt{1+4a}]$. Thus

$$\sqrt{2 + \sqrt{2 + \sqrt{2 + \cdots}}} = 2 \text{ and } \sqrt{1 + \sqrt{1 + \sqrt{1 + \cdots}}} = \varphi.$$

Chapter 6

6.1 See Figure S6.1.

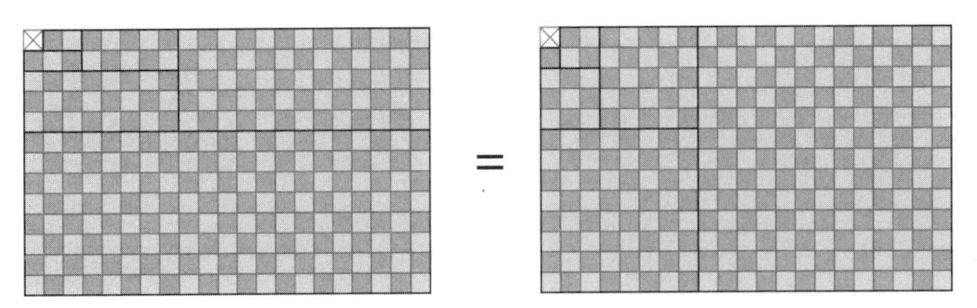

FIGURE S6.1

6.2 See Figure S6.2

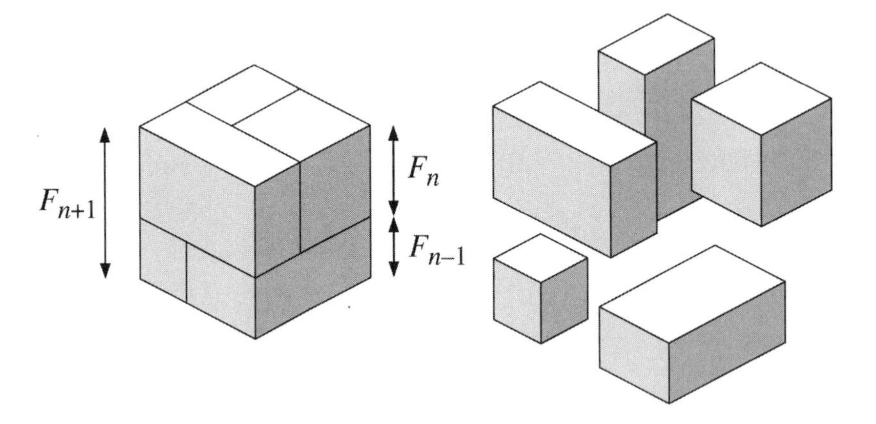

FIGURE S6.2

6.3 The identities in (6.12) yield $5F_n = 2L_{n+1} - L_n = L_{n+1} + L_{n-1}$ and $L_n = 2F_{n+1} - F_n = F_{n+1} + F_{n-1}$.

6.4 (a) $\sum_{k=1}^{n} L_{2k-1} = \sum_{k=1}^{n} (L_{2k} - L_{2k-2}) = L_{2n} - L_0 = L_{2n} - 2$.
(b) $\sum_{k=1}^{n} L_{2k} = \sum_{k=1}^{n} (L_{2k+1} - L_{2k-1}) = L_{2n+1} - L_1 = L_{2n+1} - 1$.

6.5 (a) Solve (b) first, then set $k = n$.
(b)

$$\begin{aligned} F_k L_n + F_n L_k &= F_k (F_{n-1} + F_{n+1}) + F_n (F_{k-1} + F_{k+1}) \\ &= (F_k F_{n+1} + F_{k-1} F_n) + (F_n F_{k+1} + F_{n-1} F_k) \\ &= F_{n+k} + F_{n+k} = 2F_{n+k}. \end{aligned}$$

6.6 Relabel Figure 6.7 with Lucas numbers and follow the steps in the proof of Theorem 6.5 to reach $L_{n+1}L_{k+1} + L_n L_k = L_2 L_{n+k} + L_1 L_{n+k-1} = 3L_{n+k} + L_{n+k-1} = 2L_{n+k} + L_{n+k+1} = L_{n+k} + L_{n+k+2}$, and note that $L_{n+k} + L_{n+k+2} = 5F_{n+k+1}$ (from Exercise 6.3).

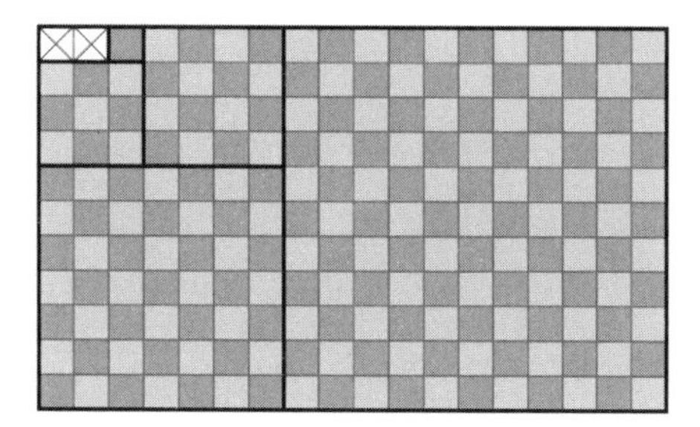

FIGURE S6.3

6.7 See Figure S6.3 [Brousseau, 1972] .

6.8 In the quadratic equation $x^2 - bx + c = 0$, the coefficient b is the sum of the roots, and the coefficient c is the product. Thus a quadratic equation with roots φ^n and $(-\varphi)^{-n}$ is $x^2 - L_n x + (-1)^n = 0$.

6.9 Since F_{3k} is even, both F_{3k-1} and F_{3k+1} are odd. Hence $L_{3k} = F_{3k-1} + F_{3k+1}$ is even (see Exercise 6.3).

6.10 (a) See the second identity in (6.12).
(b) The harmonic mean of a and b is $2ab/(a+b)$, so from (6.12) and Exercise 6.5(a) the harmonic mean of F_n and L_n is $2F_{2n}/2F_{n+1}$.

Chapter 7

7.1 From the proof of Theorem 7.4 we have $N_p = T_{3n+1}$ and it is easily verified that $T_{3n+1} = T_{2n+1} + T_{2n} + T_n$. See Figure S7.1 for a visual proof.

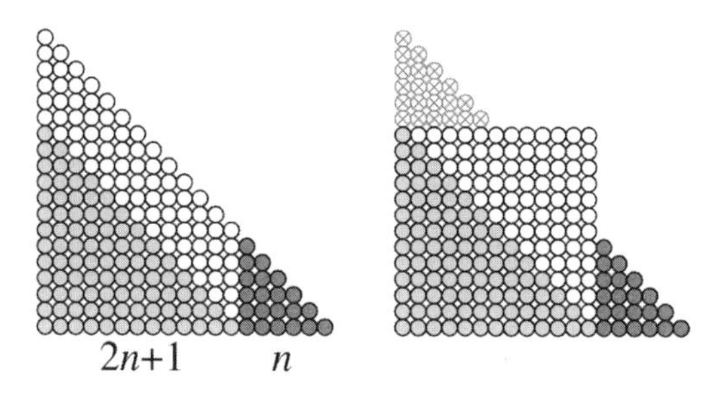

FIGURE S7.1

7.2 The pattern is $N_p = T_{2^p-1} = 3T_{2^{p-1}-1} + T_{2^{p-1}}$. So we need only show that $T_{2n-1} = 3T_{n-1} + T_n$ (which can be verified by simple algebra) and set $n = 2^{p-1}$. In Figure S7.2 we illustrate $T_{2n-1} = 3T_{n-1} + T_n$ for $n = 12$, i.e., T_{23} partitioned into three copies of T_{11} (in light gray) and one T_{12} (in dark gray).

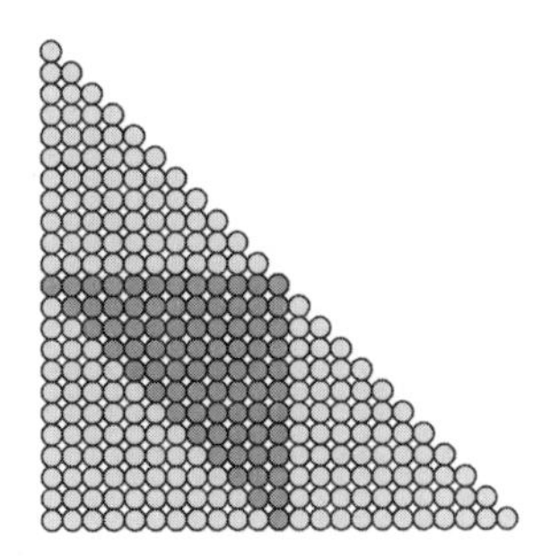

FIGURE S7.2

7.3 For $p = 2k+1$ the pattern is $N_p = T_{2^{2k+1}-1} = T_{2^k-1}T_{2^{k+1}} + T_{2^k}T_{2^{k+1}-1}$. So we need only show that $T_{mn-1} = T_{m-1}T_n + T_mT_{n-1}$ (which can be verified by simple algebra) and set $m = 2^k$, $n = 2^{k+1}$. In Figure S7.3 we illustrate $T_{mn-1} = T_{m-1}T_n + T_mT_{n-1}$ for $(m,n) = (4,6)$, i.e., T_{23} partitioned into T_3 copies of T_6 (in dark gray) and T_4 copies of T_5 (in light gray).

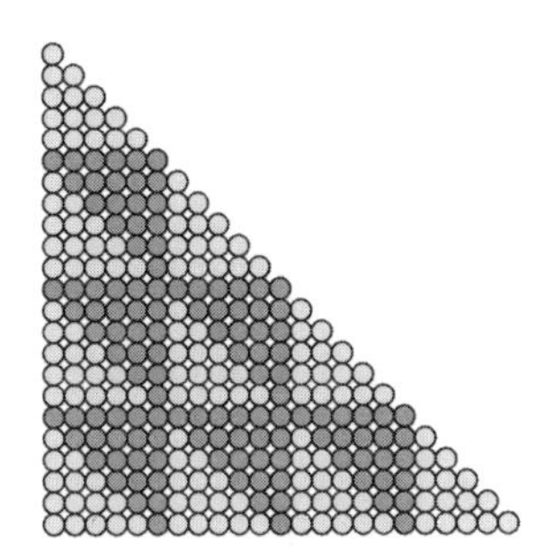

FIGURE S7.3

7.4 From the proof of Theorem 7.4, $N_p \equiv 1 \pmod 3$, and $N_p \equiv 0 \pmod 4$ since N_p is divisible by 4. Solving these two congruences simultaneously (as in Example 3.3) yields $N_p \equiv 4 \pmod{12}$.

7.5 For $p \geq 5$ we have $N_p \equiv 0 \pmod{16}$ and $N_p \equiv 1 \pmod 9$. Solving these two congruences simultaneously (as in Example 3.3) yields $N_p \equiv 64 \pmod{144}$ so that $N_p = 144k + 64$ for some positive integer k. Hence in base 12 the rightmost two digits of N_p are 54, the base 12 number equal to 64 in base 10.

7.6 From the representation of N_p as the sum of a geometric progression in Section 7.2, we have $N_p = 111\cdots100\cdots0_{\text{base }2}$, with p ones followed by $p-1$ zeros.

7.7 $\sqrt{1+8N_p} = 2^{p+1} - 1$ since $1 + 8 \cdot 2^{p-1}(2^p - 1) = (2^{p+1} - 1)^2$. For a visual proof, recall that N_p is triangular and see Figure 2.4.

7.8 When $p \geq 3$, N_p is a multiple of 4. Hence when the units digit is 6, the rightmost two digits must be 16, 36, 56, 76, or 96. All five possibilities are known to occur.

7.9 (a) It suffices to show that

$$2^p \equiv 2 \pmod{6} \text{ and } 2^p \equiv 8 \pmod{24}.$$

As in the proof of Theorem 7.4 we have $2^p \equiv (-1)^p \equiv -1 \equiv 2 \pmod{3}$. Using the Chinese remainder theorem to solve $2^p \equiv 0 \pmod{2}$ and $2^p \equiv 2 \pmod{3}$ yields $2^p \equiv 2 \pmod{6}$; similarly solving $2^p \equiv 0 \pmod{8}$ and $2^p \equiv 2 \pmod{3}$ yields $2^p \equiv 8 \pmod{24}$.

(b) If $p = 4k + 1$, then $2^p = 2 \cdot 4^{2k} \equiv 2 \cdot (-1)^{2k} \equiv 2 \pmod{5}$, so that 2^p ends in 2 or 7 (base 10). Since 2^p is even it ends in 2, and hence $M_p = 2^p - 1$ ends in 1. If $p = 4k + 3$, then $2^p = 8 \cdot 4^{2k} \equiv 3 \cdot (-1)^{2k} \equiv 3 \pmod{5}$, so that 2^p ends in 3 or 8 (base 10). Since 2^p is even it ends in 8, and hence $M_p = 2^p - 1$ ends in 7.

7.10 No, since $2^p - 1 \not\equiv 1 \pmod{4}$.

Bibliography

Andreescu, T., D. Andrica, and I. Cucurezeanu. 2010. *An introduction to Diophantine equations: A problem-based approach*, Birkhäuser Verlag, New York. MR2723590

Andrews, G. E. 1971. *Number theory*, W. B. Saunders Co., Philadelphia, Pa.-London-Toronto, Ont. MR0309838

Apostol, T. 2000. *Irrationality of the square root of two—A geometric proof*, Amer. Math. Monthly **107**, 841–842.

Bell, E. T. 1990. *The last problem*, 2nd ed., MAA Spectrum, Mathematical Association of America, Washington, DC. With an introduction and notes by Underwood Dudley; With a chapter by D. H. Lehmer. MR1075993

Bloom, D. M. 1995. *A one-sentence proof that $\sqrt{2}$ is irrational*, Math. Magazine **68**, no. 4, 286. MR1573107

Boardman, M. 2000. *Proof without words: Pythagorean runs*, Math. Magazine **73**, no. 1, 59. MR1573436

Brousseau, A. 1972. *Fibonacci numbers and geometry*, Fibonacci Quarterly, **10**, 146–158.

Carmichael, R. D. 1914. *The Theory of Numbers*, John Wiley & Sons, New York.

Conway, J. H. 2005. "The power of mathematics", in A. F. Blackwell and D. J. C. MacKay, *Power*, Cambridge University Press, Cambridge, 36–50.

Deza, E. and M. M. Deza. 2012. *Figurate numbers*, World Scientific, Singapore. MR2895686

Edgar, T. 2017. *Proof without words: A recursion for triangular numbers and more*, Math. Magazine **90**, no. 2, 124–125. MR3626285

Frenzen, C. L. 1997. *Proof without words: sums of consecutive positive integers*, Math. Magazine **70**, no. 4, 294. MR1573264

Gardner, M. 1973. *Mathematical games*, Scientific American **229**, 115.

Gardner, M. 2001. *The colossal book of mathematics*, W. W. Norton & Co., New York.

Goldberg, D., Personal communication, 1993.

Golomb, S. W. 1965. *Classroom notes: Combinatorial proof of Fermat's "Little" Theorem*, Amer. Math. Monthly **63**, no. 10, 718. MR1529479

Golomb, S. W. 1965. *A geometric proof of a famous identity*, Math. Gazette **49**, 198–200.

Gomez, J. 2005. *Proof without words: Pythagorean triples and factorizations of even squares*, Math. Magazine **78**, 14.

Hall, A. 1970. *Genealogy of Pythagorean triads*, Math. Gazette **54**, 377–379.

Heath, T. L. 1956. *The Thirteen Books of Euclid's Elements*, Dover Publications, Inc., New York.

Jones, C. and N. Lord 1999. *Characterizing non-trapezoidal numbers*, Math. Gazette **83**, 262–263.

Kalman, D. 1991. $(1 + 2 + \cdots + n)(2n + 1) = 3(1^2 + 2^2 + \cdots n^2)$, College Math. J. **22**, 124.

Koshy, T. 2001. *Fibonacci and Lucas numbers with applications*, Pure and Applied Mathematics (New York), Wiley-Interscience, New York. MR1855020

Křížek, M., F. Luca, and L. Somer. 2001. *17 lectures on Fermat numbers: From number theory to geometry*, CMS Books in Mathematics/Ouvrages de Mathématiques de la SMC, vol. 9, Springer-Verlag, New York. With a foreword by Alena Šolcová. MR1866957

Larson, L. 1985. *A discrete look at* $1 + 2 + \cdots + n$, College Math. J. **16**, 369–382.

Marshall, D. C., E. Odell, and M. Starbird. 2007. *Number theory through inquiry*, MAA Textbooks, Mathematical Association of America, Washington, DC. MR2501633

Miller, S. J. and D. Montague. 2012. *Picturing irrationality*, Math. Magazine **85**, no. 2, 110–114. MR2910300

Miller, W. A. 1993. *Proof without words: Sum of pentagonal numbers*, Math. Magazine **66**, no. 5, 325. MR1572989

Nelsen, R. B. and H. Unal. 2012. *Proof without words: Runs of triangular numbers*, Math. Magazine **85**, no. 5, 373. MR3287892

Ollerton, R. L. 2008. *Proof without words: Fibonacci tiles*, Math. Magazine **81**, 302.

Price, L. 2011 *The Pythagorean tree: A new species*, `http://arxiv.org/pdf/0809.4324.pdf`.

Ribenboim, P. 1996. *The new book of prime number records*, Springer-Verlag, New York. MR1377060

Schub, P. 1950. *A minor Fibonacci curiosity*, Scripta Mathematica, **16**, no. 3, 214.

Sierpiński, W. 2003. *Pythagorean triangles*, Dover Publications, Inc. Mineola, NY.

Spira, R. 1962. *The Diophantine equation* $x^2 + y^2 + z^2 = m^2$, Amer. Math. Monthly **69**, 360–364. MR0139574

Wall, R. 1984. *Even perfect numbers* mod 7, Fibonacci Quarterly **22**, 274–275.

Walser, H. 2001. *The golden section*, MAA Spectrum, Mathematical Association of America, Washington, DC. Translated from the second (1996) German edition by Peter Hilton with the assistance of Jean Pedersen; With a foreword by Hilton. MR1843862

Walser, H. 2011. *Proof without words: Fibonacci trapezoids*, Math. Magazine **84**, 295.

Zerger, M. J. 1990. *Proof without words: Sums of triangular numbers*, Math. Magazine **63**, no. 5, 314. MR1572828

Index